BIRDS AND MAN

W. H. HUDSON

ZINC READ

Published by
This edition copyright © 2024 by Zinc Read
Design and setting By Zinc Read
Email- zincread@gmail.com
ISBN-978-93-6277-828-4

CONTENTS

CHAPTER I

BIRDS AT THEIR BEST

By Way of Introduction Years ago, in a chapter concerning eyes in a book of Patagonian memories, I spoke of the unpleasant sensations produced in me by the sight of stuffed birds. Not bird skins in the drawers of a cabinet, it will be understood, these being indispensable to the ornithologist, and very useful to the larger class of persons who without being ornithologists yet take an intelligent interest in birds. The unpleasantness was at the sight of skins stuffed with wool and set up on their legs in imitation of the living bird, sometimes (oh, mockery!) in their "natural surroundings."

These "surroundings" are as a rule constructed or composed of a few handfuls of earth to form the floor of the glass case--sand, rock, clay, chalk, or gravel; whatever the material may be it invariably has, like all "matter out of place," a grimy and depressing appearance. On the floor are planted grasses, sedges, and miniature bushes, made of tin or zinc and then dipped in a bucket of green paint. In the chapter referred to it was said, "When the eye closes in death, the bird, except to the naturalist, becomes a mere bundle of dead feathers; crystal globes may be put into the empty sockets, and a bold life-imitating attitude given to the stuffed specimen, but the vitreous orbs shoot forth no life-like glances: the 'passion and the life whose fountains are within' have vanished, and the best work of the taxidermist, who has given a life to his bastard art, produces in the mind only sensations of irritation and disgust."

That, in the last clause, was wrongly writ. It should have been *my* mind, and the minds of those who, knowing living birds intimately as I do, have the same feeling about them.

This, then, being my feeling about stuffed birds, set up in their "natural surroundings," I very naturally avoid the places where they are exhibited. At Brighton, for instance, on many occasions when I have visited and stayed in that town, there was no inclination to see the Booth Collection, which is supposed to be an ideal collection of British birds; and we know it was the life-work of a zealous ornithologist who was also a wealthy man, and who

spared no pains to make it perfect of its kind. About eighteen months ago I passed a night in the house of a friend close to the Dyke Road, and next morning, having a couple of hours to get rid of, I strolled into the museum. It was painfully disappointing, for though no actual pleasure had been expected, the distress experienced was more than I had bargained for. It happened that a short time before,

I had been watching the living Dartford warbler, at a time when the sight of this small elusive creature is loveliest, for not only was the bird in his brightest feathers, but his surroundings were then most perfect--

The whin was frankincense and flame.

His appearance, as I saw him then and on many other occasions in the furze-flowering season, is fully described in a chapter in this book; but on this particular occasion while watching my bird I saw it in a new and unexpected aspect, and in my surprise and delight I exclaimed mentally, "Now I have seen the furze wren at his very best!"

It was perhaps a very rare thing--one of those effects of light on plumage which we are accustomed to see in birds that have glossed metallic feathers, and, more rarely, in other kinds. Thus the turtle-dove when flying from the spectator with a strong

sunlight on its upper plumage, sometimes at a distance of two to three hundred yards, appears of a

shining whiteness.

I had been watching the birds for a couple of hours, sitting quite still on a tuft of heather among the furze-bushes, and at intervals they came to me, impelled by curiosity and solicitude, their nests being near, but, ever restless, they would never remain more than a few seconds at a time in sight.

The prettiest and the boldest was a male, and it was this bird that in the end flew to a bush within twelve yards of where I sat, and perching on a spray about on a level with my eyes exhibited himself to me in his characteristic manner, the long tail raised, crest erect, crimson eye sparkling, and throat puffed out with his little scolding notes. But his colour was no longer that of the furze wren: seen at a distance the upper plumage always appears slaty-black; near at hand it is of a deep slaty-brown; now it was dark, sprinkled or frosted over with a delicate greyish-white, the white of oxidised silver; and this rare and beautiful appearance continued for a space of about twenty seconds; but no sooner did he flit to another spray than it vanished, and he was once more the slaty-brown little bird with a chestnut-red breast.

It is unlikely that I shall ever again see the furze wren in this aspect, with a curious splendour wrought by the sunlight in the dark but semi-translucent delicate feathers of his mantle; but its

image is in the mind, and, with a thousand others equally beautiful, remains to me a permanent possession.

As I went in to see the famous Booth Collection, a thought of the bird I have just described came into my mind; and glancing round the big long room with shelves crowded with stuffed birds, like the crowded shelves of a shop, to see where the Dartford warblers were, I went straight to the case and saw a group of them fastened to a furze-bush, the specimens twisted by the stuffer into a variety of attitudes--ancient, dusty, dead little birds, painful to look at--a libel on nature and an insult to a man's intelligence.

It was a relief to go from this case to the others, which were not of the same degree of badness, but all, like the furze wrens, were in their natural surroundings--the pebbles, bit of turf, painted leaves,

and what not, and, finally, a view of the wide world beyond, the green earth and the blue sky, all painted on the little square of deal or canvas which formed the back of the glass case.

Listening to the talk of other visitors who were making the round of the room, I heard many sincere expressions of admiration: they were really pleased and thought it all very wonderful. That is, in fact, the common feeling which most persons express in such places, and, assuming that it is sincere, the obvious explanation is that they know no better.

They have never properly seen anything in nature, but have looked always with mind and the inner vision preoccupied with other and familiar things--indoor scenes and objects, and scenes described in books. If they had ever looked at wild birds properly--that is to say, emotionally--the images of such sights would have remained in their minds; and, with such a standard for comparison, these dreary remnants of dead things set before them as restorations and as semblances of life would have only produced a profoundly depressing effect.

We hear of the educational value of such exhibitions, and it may be conceded that they might be

made useful to young students of zoology, by distributing the specimens over a large area, arranged

in scattered groups so as to give a rough idea of the relationship existing among its members, and of all together to other neighbouring groups, and to others still further removed. The one advantage of such a plan to the young student would be, that it would help him to get rid of the false notion, which classification studied in books invariably produces, that

nature marshals her species in a line or row, or her genera in a chain. But no such plan is ever attempted, probably because it would only be for the benefit of about one person in five hundred visitors, and the expense would be too great.

As things are, these collections help no one, and their effect is confusing and in many ways injurious to the mind, especially to the young. A multitude of specimens are brought before the sight, each and every one a falsification and degradation of nature, and the impression left is of an assemblage, or mob, of incongruous forms, and of a confusion of colours.

The one comfort is that nature, wiser than our masters, sets herself against this rude system of overloading the brain. She is kind to her wild children

in their intemperance, and is able to relieve the congested mind, too, from this burden. These objects in a museum are not and cannot be viewed emotionally, as we view living forms and all nature; hence they do not, and we being what we are, cannot, register lasting impressions.

It needed a long walk on the downs to get myself once more in tune with the outdoor world after that distuning experience; but just before quitting the house in the Dyke Road an old memory came to me and gave me some relief, inasmuch as it caused me to smile. It was a memory of a tale of the Age of Fools, which I heard long years ago in the days of my youth.

I was at a small riverine port of the Plata river, called Ensenada de Barragan, assisting a friend to ship a number of sheep which he had purchased in Buenos Ayres and was sending to the Banda Oriental--the little republic on the east side of the great sea-like river. The sheep, numbering about six thousand, were penned at the side of the creek where the small sailing ships were lying close to the bank, and a gang of eight men were engaged in carrying the animals on board, taking them one by one on their backs over a narrow plank, while I stood by keeping count. The men were gauchos, all but one--a

short, rather grotesque-looking Portuguese with one eye. This fellow was the life and soul of the gang, and with his jokes and antics kept the others in a merry humour. It was an excessively hot day, and at intervals of about an hour the men would knock off work, and, squatting on the muddy bank, rest and smoke their cigarettes; and on each occasion the funny one-eyed Portuguese would relate

some entertaining history. One of these histories was about the Age of Fools, and amused me so much that I remember it to this day. It was the history of a man of that remote age, who was born out of his time, and who grew tired of the monotony of his life, even of the society of his wife, who was no whit wiser than the other inhabitants of the village they lived in. And at last he resolved to go forth and see the world, and bidding his wife and friends farewell he set out on his travels. He travelled far and met with many strange and entertaining adventures, which I must be pardoned for not relating, as this is not a story-book. In the end he returned safe and sound to his home, a much richer man than when he started; and opening his pack he spread out before his wife an immense number of gold coins, with scores of precious stones, and trinkets of the greatest value. At the sight of this glittering treasure she uttered a great scream of joy and jumping up rushed from the room. Seeing that she did not return, he went to look for her, and after some searching discovered that she had rushed down to the wine-cellar

and knocking open a large cask of wine had jumped into it and drowned herself for pure joy.

"Thus happily ended his adventures," concluded

the one-eyed cynic, and they all got up and resumed their work of carrying sheep to the boat.

It was one of the adventures met with by the man of the tale in his travels that came into my mind when I was in the Booth Museum, and caused me to smile. In his wanderings in a thinly settled district, he arrived at a village where, passing by the church, his attention was attracted by a curious spectacle. The church was a big building with a rounded roof, and great blank windowless walls, and the only door he could see was no larger than the door of a cottage. From this door as he looked a small old man came out with a large empty sack in his hands. He was very old, bowed and bent with infirmities, and his long hair and beard were white as snow. Toddling out to the middle of the churchyard he stood still, and grasping the empty sack by its top, held it open between his outstretched arms for a space of about five minutes; then with a sudden movement of his hands he closed the sack's mouth, and still grasping it tightly, hurried back to the church as fast as his stiff joints would let him, and disappeared within the door. By and by he came forth again and repeated the performance, and then again, until the traveller approached and asked him what he was doing. "I am lighting the church," said the old man; and he then went on to explain that it was a large and a fine church, full of rich ornaments, but very dark inside--so dark that when people came to service the greatest confusion prevailed, and they could not see each other or the priest, nor the priest them. It had always been so, he continued, and it was a great mystery; he had been engaged by the fathers of the village a long time back, when he was a young man, to carry sunlight in to light the interior; but though he had grown old at his task, and had carried in many, many thousands of sackfuls of sunlight every year, it still remained dark, and no one could say why it was so.

It is not necessary to relate the sequel: the reader knows by now that in the end the dark church was filled with light, that the traveller was feasted and honoured by all the people of the village, and that he left them loaded with gifts.

Parables of this kind as a rule can have no moral or hidden meaning in an age so enlightened as this; yet oddly enough we do find among us a delusion resembling that of the villagers who thought they could convey sunshine in a sack to light their dark church. It is one of a group or family of indoor delusions and illusions, which Mr Sully has not mentioned in his book on that fascinating subject.

One example of the particular delusion I have been speaking of, in which it is seen in its crudest form, may be given here.

A man walking by the water-side sees by chance a kingfisher fly past, its colour a wonderful blue, far surpassing in beauty and brilliancy any blue he has ever seen in sky or water, or in flower or stone, or any other thing. No sooner has he seen than he wishes to become the possessor of that rare loveliness, that shining object which, he fondly imagines, will be a continual delight to him and to all in his house,--an ornament comparable to that splendid stone which the poor fisherman found in a fish's belly, which was his children's plaything by day and his candle by night. Forthwith he gets his gun and shoots it, and has it stuffed and put in a glass case.

But it is no longer the same thing: the image of the living sunlit bird flashing past him is in his mind and creates a kind of illusion when he looks at his feathered mummy, but the lustre is not visible to others.

It is because of the commonness of this delusion that stuffed kingfishers, and other brilliant species, are to be seen in the parlours of tens of thousands of cottages all over the land. Nor is it only those who live in cottages that make this mistake; those who care to look for it will find that it exists in some degree in most minds--the curious delusion that the lustre which we see and admire is in the case, the coil, the substance which may be grasped, and not in the spirit of life which is within and the atmosphere and miracle-working sunlight which are without.

To return to my own taste and feelings, since in the present chapter I must be allowed to write on Man (myself to wit) and Birds, the other chapters being occupied with the subject of Birds and Man.

It has always, or since I can remember, been my ambition and principal delight to see and hear every bird at its best. This is here a comparative term, and simply means an unusually attractive aspect of the bird, or a very much better than the ordinary one. This may result from a fortunate conjunction of circumstances, or may be due to a peculiar harmony between the creature and its surroundings; or in some instances, as in that given above of the Dartford warbler, to a rare effect of the sun.

In still other cases, motions and antics, rarely seen, singularly graceful, or even grotesque, may give the best impression. After one such impression has been received, another equally excellent may follow at a later date: in that case the second impression does not obliterate, or is not superimposed upon the former one; both remain as permanent possessions of the mind, and we may thus have several mental pictures of the same species.

It is the same with all minds with regard to the objects and scenes which happen to be of special interest. The following illustration will serve to make the matter clearer to readers who are not accustomed to pay attention to their own mental processes. When any common object, such as a chair, or spade, or apple, is thought of or spoken of, an image of a picture of it instantly comes before the mind's eye; not of a particular spade or apple, but of a type representing the object which exists in the mind ready for use on all occasions. With the question of the origin of this type, this spade or apple of the mind, we need not concern ourselves here. If the object thought or spoken of be an animal--a horse let us say, the image seen in the mind will in most cases be as in the foregoing case a type existing in the mind and not of an individual.

But if a person is keenly interested in horses generally, and is a rider and has owned and loved many horses, the image of some particular one which he has known or has looked at with appreciative eyes will come to mind; and he will also be able to call up the images of dozens or of scores of horses he has known or seen in the same way. If on the other hand we think of a rat, we see not any individual but a type, because we have no interest in or no special feeling with regard to such a creature, and all the successive images we receive of it become merged in one--the type which already existed in the mind and was probably formed very early in life. With the dog for subject the case is different: dogs are more with us--we know them intimately and have perhaps regarded many individuals with affection; hence the image that rises in the mind is as a rule of some dog we have known.

The important point to be noted is, that while each and everything we see registers an impression in the brain, and may be recalled several minutes, or hours, or even days afterwards, the only permanent impressions are of the sights which we have viewed emotionally. We may remember that we have seen a thousand things in which at some later period an interest has been born in the mind, when it would be greatly to our pleasure and even profit to recover their images, and we strive and ransack our brains to do so, but all in vain: they have been lost for ever because we happened not to be interested in the originals, but viewed them with indifference, or unemotionally.

With regard to birds, I see them mentally in two ways: each species which I have known and observed in its wild state has its type in the mind--an image which I invariably see when I think of the species; and, in addition, one or two or several, in some cases as many as fifty, images of the same species of bird as it appeared at some exceptionally favourable moment and was viewed with peculiar interest and pleasure.

Of hundreds of such enduring images of our commonest species I will here describe one before concluding with this part of the subject.

The long-tailed or bottle-tit is one of the most delicately pretty of our small woodland birds, and among my treasures, in my invisible and intangible album, there were several pictures of him which I had thought unsurpassable, until on a day two years ago when a new and better one was garnered. I was walking a few miles from Bath by the Avon where it is not more than thirty or forty yards wide, on a cold, windy, very bright day in February. The opposite bank was lined with bushes growing close to the water, the roots and lower trunks of many of them being submerged, as the river was very full; and behind this low growth the ground rose abruptly, forming a long green hill crowned with tall beeches.

I stopped to admire one of the bushes across the stream, and I wish I could now say what its species was: it was low with widespread branches close to the surface of the water, and its leafless twigs were adorned with catkins resembling those of the black poplar, as long as a man's little finger, of a rich dark-red or maroon colour. A party of about a dozen

long-tailed tits were travelling, or drifting, in their usual desultory way, through the line of bushes towards this point, and in due time they arrived, one by one, at the bush I was watching, and finding it sheltered from the wind they elected to remain at that spot. For a space of fifteen minutes I looked on with delight, rejoicing at the rare chance which had brought that exquisite bird-and plant-scene before me. The long deep-red pendent catkins and the little pale birdlings among them in their grey and rose-coloured plumage, with long graceful tails and minute round, parroty heads; some quietly perched just above the water, others moving

about here and there, occasionally suspending themselves back downwards from the slender terminal twigs--the whole mirrored below. That magical effect of water and sunlight gave to the scene a somewhat fairy-like, an almost illusory, character.

Such scenes live in their loveliness only for him who has seen and harvested them: they cannot be pictured forth to another by words, nor with the painter's brush, though it be charged with *tintas orientales*; least of all by photography, which brings all things down to one flat, monotonous, colourless shadow of things, weary to look at.

From sights we pass to the consideration of

sounds, and it is unfortunate that the two subjects have to be treated consecutively instead of together, since with birds they are more intimately joined than in any other order of beings; and in images of bird life at its best they sometimes cannot be dissociated;--the aerial form of the creature, its

harmonious, delicate tints, and its grace of motion; and the voice, which, loud or low, is aerial too, in harmony with the form.

We know that as with sights so it is with sounds: those to which we listen attentively, appreciatively, or in any way emotionally, live in the mind, to be recalled and reheard at will. There is no doubt that in a large majority of persons this retentive power is far less strong with regard to sounds than sights, but we are all supposed to have it in some degree.

So far, I have met with but one person, a lady, who is without it: sounds, in her case, do not register an impression in the brain, so that with regard to this sense she is in the condition of civilised man generally with regard to smells. I say of civilised man, being convinced that this power has become obsolete in us, although it appears to exist in savages and in the lower animals. The most common

sounds, natural or artificial, the most familiar bird-notes, the lowing of a cow, the voices of her nearest and dearest friends, and simplest melodies sung or played, cannot be reproduced in her brain: she remembers them as agreeable sounds, just as we all remember that certain flowers and herbs have agreeable odours; but she does not *hear* them. Probably there are not many persons in the same case; but in such matters it is hard to know what the real condition of another's mind may be. Our acquaintances

refuse to analyse or turn themselves inside

out merely to gratify a curiosity which they may think idle. In some cases they perhaps have a kind of superstition about such things: the secret processes of their mind are *their* secret, or "business,"

and, like the secret and *real* name of a person among some savage tribes, not to be revealed but at the risk of giving to another a mysterious power over their lives and fortunes. Even worse than the reticent, the superstitious, and the simply unintelligent, is the highly imaginative person who is only too ready to answer all inquiries, who catches at what you say in explanation, divines what you want, and instantly (and unconsciously) invents something to tell you.

But we may, I think, take it for granted that the faculty of retaining sounds is as universal as that of retaining sights, although, speaking generally, the impressions of sounds are less perfect and lasting than those which relate to the higher, more intellectual sense of vision; also that this power varies greatly in different persons. Furthermore, we see in the case of musical composers, and probably of most musicians who are devoted to their art, that this faculty is capable of being trained and developed to an extraordinary degree of efficiency. The composer sitting pen in hand to write his score in his silent room hears the voices and the various instruments, the solos and orchestral sounds, which are

in his thoughts. It is true that he is a creator, and listens mentally to compositions that have never been previously heard; but he cannot imagine, or cannot *hear* mentally, any note or combination of notes which he has never heard with his physical sense. In creating he selects from the infinite variety of sounds whose images exist in his mind, and, rearranging them, produces new effects.

The difference in the brains, with regard to their sound-storing power, of the accomplished musician and the ordinary person who does not know one tune from another and has but fleeting impressions of sounds in general, is no doubt enormous; probably it is as great as that which exists in the logical faculty between a professor of that science in one of the Universities and a native of the Andaman Islands or of Tierra del Fuego. It is, we see, a question of training: any person with a normal brain

who is accustomed to listen appreciatively to certain sounds, natural or artificial, must store his mind with the images of such sounds. And the open-air naturalist, who is keenly interested in the language of birds, and has listened with delight to a great variety of species, should be as rich in such impressions as the musician is with regard to musical sounds. Unconsciously he has all his life been training the faculty.

With regard to the durability of the images, it may be thought by some that, speaking of birds, only those which are revived and restored, so to speak, from time to time by fresh sense-impressions remain permanently distinct. That would naturally be the first conclusion most persons would arrive at, considering that the sound-images which exist in their minds are of the species found in their own country, which they are able to hear occasionally, even if at very long intervals in some cases. My own experience proves that it is not so; that a man may cut himself off from the bird life he knows, to make his home in another region of the globe thousands of miles away, and after a period exceeding a quarter of a century, during which he has become intimate with a wholly different bird life, to find that the old sound-images, which have never been refreshed with new sense-impressions, are as distinct as they ever were, and seem indeed imperishable.

I confess that, when I think of it, I am astonished myself at such an experience, and to some it must seem almost incredible. It will be said, perhaps, that in the infinite variety of bird-sounds heard anywhere there must be innumerable notes which closely resemble, or are similar to, those of other species in other lands, and, although heard in a different order, the old images of cries and calls and songs are thus indirectly refreshed and kept alive.

I do not think that has been any real help to me.

Thus, I think of some species which has not been thought of for years, and its language comes back at call to my mind. I listen mentally to its various notes, and there is not one in the least like the notes of any British species. These images have therefore never received refreshment. Again, where there is a resemblance, as in the trisyllabic cry of the common sandpiper and another species, I listen mentally to one, then to the other, heard so long ago, and hear both distinctly, and comparing the two, find a considerable difference, one being a thinner, shriller, and less musical sound than the other. Still again, in the case of the blackbird, which has a considerable variety in its language, there is one little chirp familiar to every one--a small round drop of sound of a musical, bell-like character. Now it happens that one of the true thrushes of South America, a bird resembling our song-thrush, has an almost identical bell-like chirp, and so far as that small drop of sound is concerned the old image may be refreshed by new sense-impressions.

Or I might even say that the original

image has been covered by the later one, as in the case of the laughter-like cries of the Dominican and the black-backed gulls. But with regard to the thrushes, excepting that small drop of sound, the language of the two species is utterly different.

Each has a melody perfect of its kind: the song of the foreign bird is not fluty nor mellow nor placid like that of the blackbird, but has in a high degree that quality of plaintiveness and gladness commingled which we admire in some fresh and very

beautiful human voices, like that described in Lowell's lines "To Perdita Singing":--

It hath caught a touch of sadness,
Yet it is not sad;
It hath tones of clearest gladness,
Yet it is not glad.

Again, that foreign song is composed of many notes, and is poured out in a stream, as a skylark sings; and it is also singular on account of the contrast between these notes which suggest human

feeling and a purely metallic, bell-like sound, which, coming in at intervals, has the effect of the triangle in a band of wind instruments. The image of this beautiful song is as distinct in my mind as that of the blackbird which I heard every day last summer from every green place.

Doubtless there are some and perhaps a good

many ornithologists among us who have been abroad to observe the bird life of distant countries, and who when at home find that the sound-impressions they have received are not persistent, or, if not wholly lost, that they grow faint and indistinct, and become increasingly difficult to recall. They can no longer *listen* to those over-sea notes and songs as they can, mentally, to the cuckoo's call in spring, the wood-owl's hoot, to the song of the skylark and of the tree-pipit, the reeling of the night-jar and the startling scream of the woodland jay, the deep human-like tones of the raven, the inflected wild cry of the curlew, and the beautiful wild whistle of the widgeon, heard in the silence of the night on some lonely mere.

The reason is that these, and numberless more, are the sounds of the bird life of their own home and country; the living voices to which they listened when they were young and the senses keener than now, and their enthusiasm greater; they were in fact heard with an emotion which the foreign species never inspired in them, and thus heard, the images of the sounds were made imperishable.

In my case the foreign were the home birds, and on that account alone more to me than all others; yet I escaped that prejudice which the British naturalist is never wholly without--the notion that the home bird is, intrinsically, better worth listening to than the bird abroad. Finally, on coming to this country, I could not listen to the birds coldly, as an English naturalist would to those of, let us say, Queensland, or Burma, or Canada, or Patagonia,

but with an intense interest; for these were the birds which my forbears had known and listened to all their lives long; and my imagination was fired by all that had been said of their charm, not indeed by frigid ornithologists, but by a long succession of great poets, from Chaucer down to those of our own time. Hearing them thus emotionally their notes became permanently impressed on my mind, and I found myself the happy possessor of a large number of sound-images representing the bird language of two widely separated regions.

To return to the main point--the durability of the impressions both of sight and sound.

In order to get a more satisfactory idea of the number and comparative strength or vividness of the images of twenty-six years ago remaining to me after so long a time than I could by merely thinking about the subject, I drew up a list of the species of birds observed by me in the two adjoining districts of La Plata and Patagonia. Against the

name of each species the surviving sight-and sound-impressions were set down; but on going over this

first list and analysis, fresh details came to mind, and some images which had become dimmed all at once grew bright again, and to bring these in, the work had to be redone; then it was put away and the

subject left for a few days to the "subliminal consciousness,"

after which I took it up once more and

rewrote it all--list and analysis; and I think it now gives a fairly accurate account of the state of these old impressions as they exist in memory.

This has not been done solely for my own gratification.

I confess to a very strong feeling of curiosity as to the mental experience on this point of other field naturalists; and as these, or some of them, may have the same wish to look into their neighbours'

minds that I have, it may be that the example given here will be followed.

My list comprises 226 species--a large number to remember when we consider that it exceeds by about 16 or 18 the number of British species; that is to say, those which may truly be described as belonging to these islands, without including the waifs and strays and rare visitants which by a fiction are described as British birds. Of the 226, the sight-impressions of 10 have become indistinct, and one has been completely forgotten. The sight of a specimen might perhaps revive an image of this lost one as it was seen, a living wild bird; but I do not know. This leaves 215, every one of which I can mentally see as distinctly as I see in my mind the common species I am accustomed to look at

every day in England--thrush, starling, robin, *etc.*

A different story has to be told with regard to the language. To begin with, there are no fewer than 34 species of which no sound-impressions were received. These include the habitually silent kinds--the stork, which rattles its beak but makes no

vocal sound, the painted snipe, the wood ibis, and a few more; species which were rarely seen and emitted no sound--condor, Muscovy duck, harpy eagle, and others; species which were known only as winter visitants, or seen on migration, and which at such seasons were invariably silent.

Thus, those which were heard number 192. Of

these the language of 7 species has been completely forgotten, and of 31 the sound-impressions have now become indistinct in varying degrees. Deducting those whose notes have become silent and are not clearly heard in the mind, there remain 154

species which are distinctly remembered. That is to say, when I think of them and their language, the cries, calls, songs, and other sounds are reproduced in the mind.

Studying the list, in which the species are ranged in order according to their affinities, it is easy to see why the language of some, although not many, has been lost or has become more or less indistinct.

In some cases it is because there was nothing distinctive or in any way attractive in the notes; in other cases because the images have been covered and obliterated by others--the stronger images of closely-allied species. In the two American families of tyrant-birds and woodhewers, neither of which are songsters, there is in some of the closely-related species a remarkable family resemblance in their voices. Listening to their various cries and calls, the trained ear of the ornithologist can easily distinguish them and identify the species; but after

years the image of the more powerful or the better voices of, say, two or three species in a group of four or five absorb and overcome the others. I cannot find a similar case among British species to illustrate this point, unless it be that of the meadow-and rock-pipit. Strongly as the mind is impressed by the measured tinkling notes of these two songs, emitted as the birds descend to earth, it is not probable that any person who had not heard them for

a number of years would be able to distinguish or keep them separate in his mind--to hear them in their images as two distinct songs.

In the case of the good singers in that distant region, I find the voices continue remarkably dis tinct, and as an example will give the two melodious families of the finches and the troupials (Icteridae), the last an American family, related to the finches, but starling-like in appearance, many of them brilliantly coloured. Of the first I am acquainted with 12 and of the second with 14 species.

Here then are 26 highly vocal species, of which the songs, calls, chirps, and various other notes, are distinctly remembered in 23. Of the other three one was silent--a small rare migratory finch resembling the bearded-tit in its reed-loving habits, its long tail and slender shape, and partly too in its colouring.

I listened in vain for this bird's singing notes.

Of the remaining two one is a finch, the other a troupial; the first a pretty bird, in appearance a small hawfinch with its whole plumage a lovely glaucous blue; a poor singer with a low rambling song: the second a bird of the size of a starling, coloured like a golden oriole, but more brilliant; and this one has a short impetuous song composed of mixed guttural and clear notes.

Why is this rather peculiar song, of a species which on account of its colouring and pleasing social habits strongly impresses the mind, less distinct in memory than the songs of other troupials? I believe it is because it is a rare thing to hear a single song. They perch in a tree in company, like birds of paradise, and no sooner does one open his beak than all burst out together, and their singing strikes on the sense in a rising and falling tempest of confused sound. But it may be added that though

these two songs are marked "indistinct" in the list, they are not very indistinct, and become less so when I listen mentally with closed eyes.

In conclusion, it is worthy of remark that the good voices, as to quality, and the powerful ones, are not more enduring in their images than those which were listened to appreciatively for other reasons. Voices which have the quality of ventriloquism, or are in any way mysterious, or are suggestive of human tones, are extremely persistent; and such voices are found in owls, pigeons, snipe, rails, grebes, night-jars, tinamous, rheas, and in some passerine birds. Again, the swallows are not remarkable as singers compared with thrushes, finches, and other melodists; but on account of their intrinsic charm and beauty, their interesting habits, and the sentiment they inspire, we listen to them emotionally; and I accordingly find that the language of the five species of swallows I was formerly accustomed to see and hear continues as distinct in my mind as that of the chimney swallow, which I listen to every summer in England.

*

I had meant in this chapter to give three or four or half a dozen instances of birds seen at their best, instead of the one I have given --that of the long-tailed tit; and as many more images in which a

rare, unforgettable effect was produced by melody.

For as with sights so it is with sounds: for these too there are "special moments," which have

"special grace." But this chapter is already longer than it was ever meant to be, and something on another subject yet remains to be said.

The question is sometimes asked, What is the charm which you find, or say you find, in nature?

Is it real, or do these words so often repeated have a merely conventional meaning, like so many other words and phrases which men use with regard to other things? Birds, for instance: apart from the interest which the ornithologists must take in his subject, what substantial happiness can be got out of these shy creatures, mostly small and not too well seen, that fly from us when approached, and utter sounds which at their best are so poor, so thin, so trivial, compared with our soul-stirring human music?

That, briefly, is the indoor view of the subject--the view of those who, to begin with, were perhaps town-born and town-bred; who have existed amid conditions, occupied with work and pleasures, the reflex effect of which, taken altogether and in the long-run, is to dim and even deaden some of the brain's many faculties, and chiefly this best faculty of preserving impressions of nature for long years or to the end of life in all their original freshness.

Some five or six years ago I heard a speech about birds delivered by Sir Edward Grey, in which he said that the love and appreciation and study of birds was something fresher and brighter than the second-hand interests and conventional amusements in which so many in this day try to live; that the pleasure of seeing and listening to them was purer and more lasting than any pleasures of excitement, and, in the long-run, "happier than personal success."

That was a saying to stick in the mind, and

it is probable that some who listened failed to understand.

Let us imagine that in addition to this

miraculous faculty of the brain of storing innumerable brilliant images of things seen and heard, to be reproduced at call to the inner sense, there existed in a few gifted persons a correlated faculty by means of which these treasured images could be thrown at will into the mind of another; let us further imagine that some one in the audience who had wondered at that saying, finding it both dark and hard, had asked me to explain it; and that in response I had shown him, as by a swift succession of lightning flashes a score or a hundred images of birds at their best--the unimaginable loveliness, the sunlit colour, the grace of form and of motion, and the melody--how great the effect of even that brief glance into a new unknown world would have been! And if I had then said: All that you have seen--the pictures in one small room in a house of many rooms--is not after all the main thing; *that* it would be idle to speak of, since you cannot know what you do not feel, though it should be told you many times; this only can be told--the enduring images are but an incidental result of a feeling which existed already; they were never looked for, and are a free gift from nature to her worshipper;--if I had said this to him, the words of the speech which has seemed almost sheer insanity a little while before would have acquired a meaning and an appearance of truth.

*

It has curiously happened that while writing these concluding sentences some old long-forgotten lines which I read in my youth came suddenly into my mind, as if some person sitting invisible at my side and thinking them apposite to the subject had whispered them into my ear. They are lines addressed to the Merrimac River by an American

poet--whether a major or minor I do not know, having forgotten his name. In one stanza he

mentions the fact that "young Brissot" looked upon this stream in its bright flow--

And bore its image o'er the deep
To soothe a martyr's sadness,
And fresco in his troubled sleep
His prison walls with gladness.

Brissot is not generally looked upon as a "martyr"

on this side of the Atlantic, nor was he allowed to enjoy his "troubled sleep" too long after his fellow-citizens (especially the great and sea-green Incorruptible) had begun in their fraternal fashion to

thirst for his blood; but we can easily believe that during those dark days in the Bastille the image and vision of the beautiful river thousands of miles away was more to him than all his varied stores of knowledge, all his schemes for the benefit of suffering humanity, and perhaps even a better consolation than his philosophy.

It is indeed this "gladness" of old sunshine stored within us--if we have had the habit of seeing beauty everywhere and of viewing all beautiful

things with appreciation--this incalculable wealth of images of vanished scenes, which is one of our best and dearest possessions, and a joy for ever.

"What asketh man to have?" cried Chaucer,

and goes on to say in bitterest words that "now with his love" he must soon lie in "the colde grave--alone, withouten any companie."

What he asketh to have, I suppose, is a blue diamond--some unattainable good; and in the

meantime, just to go on with, certain pleasant things which perish in the using.

These same pleasant things are not to be despised, but they leave nothing for the mind in hungry days to feed upon, and can be of no comfort to one who is shut up within himself by age and bodily infirmities and the decay of the senses; on the contrary, the recollection of them at such times, as has been said, can but serve to make a present misery more poignantly felt.

It was the nobly expressed consolation of an American poet, now dead, when standing in the summer sunshine amid a fine prospect of woods and hills, to think, when he remembered the darkness of decay and the grave, that he had beheld

in nature, though but for a moment,

The brightness of the skirts of God.

CHAPTER II

BIRDS AND MAN

To most of our wild birds man must appear as a being eccentric and contradictory in his actions.

By turns he is hostile, indifferent, friendly towards them, so that they never quite know what to expect.

Take the case of a blackbird who has gradually acquired trustful habits, and builds its nest in the garden or shrubbery in sight of the friends that have fed it in frosty weather; so little does it fear that it allows them to come a dozen times a day, put the branches aside and look upon it, and even stroke its back as it sits on its eggs. By and by a neighbour's egg-hunting boy creeps in, discovers the nest, and pulls it down. The bird finds itself betrayed by its confidence; had it suspected the boy's evil intentions it would have made an outcry at his approach, as at the appearance of a cat, and the nest would perhaps have been saved. The result of such an accident would probably be the unsettling of an acquired habit, the return to the usual suspicious attitude.

Birds are able sometimes to discriminate between protectors and persecutors, but seldom very well I should imagine; they do not view the face only, but the whole form, and our frequent change of dress must make it difficult for them to distinguish the individuals they know and trust from strangers.

Even a dog is occasionally at fault when his master, last seen in black and grey suit, reappears in straw hat and flannels.

Nevertheless, if birds once come to know those who habitually protect them and form a trustful habit, this will not be abandoned on account of a little rough treatment on occasions. A lady at Worthing told me of her blackbirds breeding in her garden that they refused to be kept from the strawberries when she netted the ripening fruit.

One or more of the birds would always manage to get under the net; and when she would capture the robber and carry him, screaming, struggling and pecking at her fingers, to the end of the garden and release him, he would immediately follow her back to the bed and set himself to get at the fruit again.

In a bird's relations with other mammals there is no room for doubt or confusion; each consistently acts after its kind; once hostile, always hostile; and if once seen to be harmless, then to be trusted for ever. The fox must always be feared and detested; his disposition, like his sharp nose and red coat, is unchangeable; so, too, with the cat, stoat, weasel, *etc.* On the other hand, in the presence of herbivorous mammals, birds show no sign of suspicion; they know that all these various creatures are absolutely harmless, from the big formidable-looking bull and roaring stag, to the mild-eyed, timorous hare and rabbit. It is common to see wagtails and other species attending cattle in the pastures, and keeping close to their noses, on the look-out for the small insects driven from hiding in the grass. Daws and starlings search the backs of cattle and sheep for ticks and other parasites, and it is plain that their visits are welcome. Here a joint interest unites bird and beast; it is the nearest approach to symbiosis among the higher vertebrates of this country, but is far less advanced than the partnership which exists between the rhinoceros bird and the rhinoceros or buffalo, and between the spur-winged plover and crocodile in Africa.

One day I was walking by a meadow, adjoining the Bishop's palace at Wells, where several cows were grazing, and noticed a little beyond them a number of rooks and starlings scattered about.

Presently a flock of about forty of the cathedral jackdaws flew over me and slanted down to join the other birds, when all at once two daws dropped out of the flock on to the back of the cow standing nearest to me. Immediately five more daws followed, and the crowd of seven birds began eagerly pecking at the animal's hide. But there was not room enough for them to move freely; they pushed and struggled for a footing, throwing their wings out to keep their balance, looking like a number of hungry vultures fighting for places on a carcase; and soon two of the seven were thrown off and flew away.

The remaining five, although much straitened for room, continued for some time scrambling over the cow's back, busy with their beaks and apparently very much excited over the treasure they had discovered.

It was amusing to see how the cow took their visit; sinking her body as if about to lie down and broadening her back, and dropping her head until her nose touched the ground, she stood perfectly motionless, her tail stuck out behind like a pump-handle. At length the daws finished their feeding and quarrelling and flew away; but for some minutes the cow remained immovable in the same attitude, as if the rare and delightful sensation of so many beaks prodding and so many sharp claws scratching her hide had not yet worn off.

Deer, too, like cows, are very grateful to the daw for its services. In Savernake Forest I once witnessed a very pretty little scene. I noticed a hind lying down by herself in a grassy hollow, and as I passed her at a distance of about fifty yards it struck me as singular that she kept her head so low down that I could only see the top of it on a level with her back. Walking round to get a better sight, I saw a jackdaw standing on the turf before her, very busily pecking at her face. With my glass I was able to watch his movements very closely; he pecked round her eyes, then her nostrils, her throat, and in fact every part of her face; and just as a man when being shaved turns his face this way and that under the gentle guiding touch of the barber's fingers, and lifts up his chin to allow the razor to pass beneath it, so did the hind raise and lower and turn her face about to enable the bird to examine and reach every part with his bill. Finally the daw left the face, and, moving round, jumped on to the deer's shoulders and began a minute search in that part; having finished this he jumped on to the head and pecked at the forehead and round the bases of the ears. The pecking done, he remained for some seconds sitting perfectly still, looking very pretty with the graceful red head for a stand, the hind's long ears thrust out on either side of him. From his living perch he sprang into the air and flew away, going close to the surface; then slowly the deer raised her head and gazed after her black friend--gratefully, and regretting his departure, I could not but think.

Some birds when breeding exhibit great anxiety at the approach of any animal to the nest; but even when most excited they behave very differently towards herbivorous mammals and those which they know to be at all times the enemies of their kind. The nest of a ground-breeding species may be endangered by the proximity of a goat, sheep, deer, or any grazing animal, but the birds do not winnow the air above it, scream, make threatening dashes at its head, and try to lead it away as they would do in the case of a dog or fox. When small birds dash at and violently attack large animals and man in defence of their nest, even though the nest may not have been touched, the action appears to be purely instinctive and involuntary, almost unconscious, in fact. Acts of this kind are more often seen in humming-birds than in birds of other families; and humming-birds do not appear to discriminate between rapacious and herbivorous mammals. When they see a large animal moving about they fly close to and examine it for a few moments, then dart away; if it comes too near the nest they will attack it, or threaten an attack.

When examining their nests I have had humming-birds dash into my face. The action is similar to that of a stingless, solitary carpenter bee, common in La Plata: a round burly insect with a shining steel-blue body: when the tree or bush in which this bee has its nest is approached by a man it darts about in an eccentric manner, humming loudly, and at intervals remains suspended motionless for ten or fifteen seconds at a height of seven or eight yards above his head; suddenly it dashes quick as lightning into his face, inflicting a sharp blow. The bee falls, as if stunned, a space of a couple of feet, then rises again to repeat the action.

There is certainly a wide difference between so simple an instinctive action as this, which cannot be regarded as intelligent or conscious, and the actions of most birds in the presence of danger to their eggs or young. In species that breed on the ground in open situations the dangers to which bird and nest are exposed are of different kinds, and, leaving out the case of that anomalous creature, man, we see that as a rule the bird's judgment is not at fault. In one case it is necessary that he should guard himself while trying to save his nest; in another case the danger is to the nest only, and he then shows that he has no fear for himself. The most striking instance I have met with, bearing on this last point, relates to the action of a spur-winged lapwing observed on the Pampas. The bird's loud excited cries attracted my attention; a sheep was lying down with its nose directly over the nest, containing three eggs, and the plover was trying to make it get up and go away. It was a hot day and the sheep refused to stir; possibly the fanning of the bird's wings was grateful to her. After beating the sheep's face for some time it began pecking sharply at the nose; then the sheep raised her head, but soon grew tired of holding it up, and no sooner was it lowered than the blows and peckings began again. Again the head was raised, and lowered again with the same result, and this continued for about twelve or fourteen minutes, until the annoyance became intolerable; then the sheep raised

her head and refused to lower it any more, and in that very uncomfortable position, with her nose high in the air, she appeared determined to stay. In vain the lapwing waited, and at last began to make little jumps at the face. The nose was out of reach, but by and by, in one of its jumps, it caught the sheep's ear in its beak and remained hanging with drooping wings and dangling legs. The sheep shook her head several times and at last shook the bird off; but no sooner was it down than it jumped up and caught the ear again; then at last the sheep, fairly beaten, struggled up to her feet, throwing the bird off, and lazily walked away, shaking her head repeatedly.

How great the confidence of the plover must have been to allow it to act in such a manner!

This perfect confidence which birds have in the mammals they have been taught by experience and tradition to regard as harmless must be familiar to any one who has observed partridges associating with rabbits. The manners of the rabbit, one would imagine, must be exceedingly "upsetting" to birds of so timorous a disposition. He has a way, after a quiet interval, of leaping into activity with startling suddenness, darting violently away as if scared out of his senses; but his eccentric movements do not in the least alarm his feathered companions. One evening early in the month of March I witnessed an amusing scene near Ockley, in Surrey. I was walking towards the village about half an hour after sunset, when, hearing the loud call of a partridge, I turned my eyes in the direction of the sound and saw five birds on a slight eminence nearly in the centre of a small green field, surrounded by a low thorn hedge. They had come to that spot to roost; the calling bird was standing erect, and for some time he continued to call at intervals after the others had settled down at a distance of one or two yards apart. All at once, while I stood watching the birds there was a rustling sound in the hedge, and out of it burst two buck rabbits engaged in a frantic running fight. For some time they kept near the hedge, but fighting rabbits seldom continue long on one spot; they are incessantly on the move, although their movements are chiefly round and round now one way--flight and pursuit--then, like lightning, the foremost rabbit doubles back and there is a collision, bitings, and rolling over and over together, and in an instant they are up again, wide apart, racing like mad.

Gradually they went farther and farther from the hedge; and at length chance took them to the very spot on which the partridges had settled, and there for three or four minutes the duel went on. But the birds refused to be turned out of their quarters. The bird that had called still remained standing, expectant, with raised head, as if watching for the appearance of some loiterer, while the others all kept their places. Their quietude in the midst of that whirlwind of battle was wonderful to see. Their only movement was when one of the birds was in a direct line with a flying rabbit, when, if it stayed still, in another moment it would be struck and perhaps killed by the shock; then it would leap a few inches aside and immediately settle down again. In this way every one of the birds had been forced to move several times before the battle passed on towards the opposite side of the field and left the covey in peace.

Social animals, Herbert Spencer truly says, "take pleasure in the consciousness of one another's company;" but he appears to limit the feeling to those of the same herd, or flock, or species. Speaking of the mental processes of the cow, he tells us just how that large mammal is impressed by the sight of birds that come near it and pass across its field of vision; they are regarded in a vague way as mere shadows, or shadowy objects, flitting or blown about hither and thither over the grass or through the air.

He didn't know a cow's mind. My conviction is that all animals distinctly see in those of other species, living, sentient, intelligent beings like themselves; and that, when birds and mammals meet together, they take pleasure in the consciousness of one another's presence, in spite of the enormous difference in size, voice, habits, *etc.* I believe that this sympathy exists and is just as strong between a cow and its small volatile companion, the wagtail, as between a bird and mammal that more nearly resemble each other in size; for instance, the partridge, or pheasant, and rabbit.

The only bird with us that appears to have some such feeling of pleasure in the company of man is the robin. It is not universal, not even very common, and Macgillivray, in speaking of the confidence in men of that bird during severe weather, very truly says, "In ordinary times he is not perfectly disposed to trust in man." Any person can prove this for himself by going into a garden or shrubbery and approaching a robin. We see, too, that the bird shows intense anxiety when its nest is approached by a man; this point, however, need not be made much of, since all visitors, even its best friends, are unwelcome to the breeding bird. Still, there is no doubt that the robin is less distrustful of man than other species, but not surely because this bird is regarded by most persons with kindly feelings. The curious point is that the young birds find something in man to attract them. This is usually seen at the end of summer, when the old birds have gone into hiding, and it is then surprising to find how many of the young robins left in possession of the ground appear to take pleasure in the company of human beings. Often before a person has been many

minutes in a garden strolling about, he will discover that the quiet little spotted bird is with him, hopping and flying from twig to twig and occasionally alighting on the ground, keeping company with him,

sometimes sitting quite still a yard from his hand.

The gardener is usually attended by a friendly robin, and when he turns up the soil the bird will come down close to his feet to pick up the small grubs and worms. Is it not probable that the tameness of the tame young robin so frequently met with is, like that of the robin who keeps company with the gardener or woodman, an acquired habit; that the young bird has made the discovery that when a person is moving about among the plants, picking fruit perhaps, lurking insects are disturbed at the roots and small spiders and caterpillars shaken from the leaves? We are to the robin what the cow is to the wagtail and the sheep to the starling--a food finder.

Among the birds of the homestead the swallow is another somewhat exceptional species in his way of regarding man. He is too much a creature of the air to take any pleasure in the company of heavy animals, bound to earth; the distance is too great for sympathy to exist. When we consider how

closely he is bound and how much he is to us, it is hard to believe that he is wholly unconscious of our benefits, that when he returns in spring, overflowing with gladness, to twitter his delightful airy music round the house, he is not singing to us, glad to see us again after a long absence, to be once more our welcome guest as in past years. But so it is. When there were no houses in the land he built his nest in some rocky cavern, where a she-wolf had her lair, and his life and music were just as joyous as they are now, and the wolf suckling her cubs on the stony floor beneath was nothing to him. But if by chance she climbed a little way up or put her nose too near his nest, his lively twittering quickly changed to shrill cries of alarm and anger. And we are no more than the vanished wolf to the swallow, and so long as we refrain from peeping into his nest and handling his eggs or young, he does not know us, and is hardly conscious of our existence. All the social feelings and sympathy of the swallow are for creatures as aerial and swift-winged as itself--its playmates in the wide fields of air.

Swallows hawking after flies in a village street, where people are walking about, is a familiar sight, Swifts are just as confident. A short time ago, while standing in the churchyard at Farnham, in Surrey, watching a bunch of ten or twelve swifts racing through the air, I noticed that on each return to the church they followed the same line, doubling round the tower on the same side, then sweeping down close to the surface, and mounting again.

Going to the spot I put myself directly in their way--on their race-course as it were, at that point

where it touched the earth; but they did not on that account vary their route; each time they came back they streamed screaming past my head so near as almost to brush my face with their wings.

But I was never more struck by the unconcern at the presence of man shown by these birds--swallows, martins, and swifts--as on one occasion at Frensham, when the birds were very numerous. This was in the month of May, about five weeks after I had witnessed the fight between two rabbits, and the wonderful composure exhibited by a covey of partridges through it all. It was on a close hot morning, after a night of rain, when, walking down to Frensham Great Pond, I saw the birds hawking about near the water. The may-flies were just out, and in some mysterious way the news had been swiftly carried all over the surrounding country.

So great was the number of birds that the entire population of swallows, house-and sand-martins, and swifts, must have been gathered at that spot from the villages, farms, and sand-banks for several miles around. At the side of the pond I was approaching there is a green strip about a hundred

and twenty or a hundred and thirty yards in length and forty or fifty yards wide, and over this ground from end to end the birds were smoothly and swiftly gliding backwards and forwards. The whole place seemed alive with them. Hurrying to the spot I met with a little adventure which it may not be inapt to relate. Walking on through some scattered furze-bushes, gazing intently ahead at the swallows, I almost knocked my foot against a hen pheasant covering her young chicks on the bare ground beside a dwarf bush. Catching sight of her just in time I started back; then, with my feet about a yard from the bird, I stood and regarded her for some time. Not the slightest movement did she make; she was like a bird carved out of some beautifully variegated and highly-polished stone, but her bright round eyes had a wonderfully alert and wild expression.

With all her stillness the poor bird must
have been in an agony of terror and suspense, and I wondered how long she would endure the tension.

She stood it for about fifty seconds, then burst screaming away with such violence that her seven or eight chicks were flung in all directions to a distance of two or three feet like little balls of fluff; and going twenty yards away she dropped to the ground and began beating her wings, calling loudly.

I then walked on, and in three or four minutes was on the green ground in the thick of the swallows.

They were in hundreds, flying at various heights, but mostly low, so that I looked down on them, and they certainly formed a curious and beautiful spectacle.

So thick were they, and so straight and rapid their flight, that they formed in appearance a current, or rather many currents, flowing side by side in opposite directions; and when viewed with nearly closed eyes the birds were like black lines on the green surface. They were silent except for the occasional weak note of the sand-martin; and through it all they were perfectly regardless of me, whether I stood still or walked about among them; only when I happened to be directly in the way of a bird coming towards me he would swerve aside just far enough to avoid touching me.

In the evening of that very day the behaviour of a number of gold-crests, disturbed at my presence, surprised and puzzled me not a little; their action had a peculiar interest just then, as the encounter with the pheasant, and the sight of the multitude of swallows and their indifference towards me were still very fresh in memory. The incident has only an indirect bearing on the subject discussed here, but I think it is worth relating.

About two miles from Frensham ponds there
is a plantation of fir-trees with a good deal of gorse growing scattered about among the trees; in walking through this wood on previous occasions I had noticed that gold-crests were abundant in it. Soon after sunset on the evening in question I went through this wood, and after going about eighty to a hundred yards became conscious of a commotion of a novel kind in the branches above my head--conscious too that it had been going on for some time, and that absorbed in thought I had not remarked it. A considerable number of gold-crests were flitting through the branches and passing from tree to tree, keeping over and near me, all together uttering their most vehement cries of alarm. I stopped and listened to the little chorus of shrill squeaking sounds, and watched the birds as well as I could in the obscurity of the branches, flitting about in the greatest agitation. It was perfectly clear that I was the cause of the excitement, as the birds increased in number as long as I stood at that spot,
until there could not have been less than forty or fifty, and when I again walked on they followed.

One expects to be mobbed and screamed at by gulls, terns, lapwings, and some other species, when approaching their nesting-places, but a hostile demonstration of this kind from such minute creatures as
gold-crests, usually indifferent to man, struck me as very unusual and somewhat ridiculous. What, I asked myself, could be the reason of their sudden alarm, when my previous visits to the wood had not excited them in the least? I could only suppose that I had, without knowing it, brushed against a nest, and the alarm note of the parent birds had excited the others and caused them to gather near me, and that in the obscure light they had mistaken me for some rapacious animal. The right explanation (I think it the right one) was found by chance three months later.

In August I was in Ireland, staying at a country house among the Wicklow hills. There were several swallows' nests in the stable, one or two so low that they could be reached by the hand, and the birds went in and out regardless of the presence of any person. In a few days the young were out, sitting in rows on the roof of the house or on a low fence near it, where their parents fed them for a short time. After these young birds were able to take care of themselves they still kept about the house, and were joined by more swallows and martins from the neighbourhood. One bright sunny morning, when not fewer than two or three score of these birds were flying about the house, gaily twittering, I went into the garden to get some fruit. All at once a swallow uttered his loud shrill alarm cry overhead and at the same time darted down at me, almost grazing my hat, then mounting up he continued making swoops, screaming all the time.

Immediately all the other swallows and martins came to the spot, joining in the cry, and continued flying about over my head, but not darting at me like the first bird. For some moments I was very much astonished at the attack; then I looked round for the cat--it must be the cat, I thought.

This animal had a habit of hiding among the gooseberry bushes, and, when I stooped to pick the fruit, springing very suddenly upon my back. But pussy was nowhere near, and as the swallow continued to make dashes at me, I thought that there must be something to alarm it on my head, and at once pulled off my hat and began to examine it. In a moment the alarm cries ceased and the whole gathering of swallows dispersed in all directions. There was no doubt that my hat had caused the excitement; it was of tweed, of an obscure grey colour,

striped or barred with dark brown. Throwing it down on the ground among the bushes it struck me that its colour and markings were like those of a grey striped cat. Any one seeing it lying there would, at the first moment, have mistaken it for a cat lying curled up asleep among the bushes. Then I remembered that I had been wearing the same

delusive, dangerous-looking round tweed fishing-hat on the occasion of being mobbed by the gold-crests at Frensham. Of course the illusion could

only have been produced in a bird looking down upon the top of the hat from above.

CHAPTER III

DAWS IN THE WEST COUNTRY

Daws are more abundant in the west and south-west of England generally than in any other part of the kingdom; and they abound most in Somerset, or so it has seemed to me. It is true that the largest congregations of daws in the entire country are to be seen at Savernake in Wiltshire, where the ancient hollow beeches and oaks in the central parts of the forest supply them with all the nesting holes they require. There is no such wood of old decaying trees in Somerset to attract them to one spot in such numbers, but the country generally is singularly favourable to them. It is mainly a pastoral country with large areas of rich, low grass land, and ranges of high hills, where there are many rocky precipices such as the daw loves. For very good reasons he prefers the inland to the sea-cliff as a breeding site.

It is, to begin with, in the midst of his feeding ground, whereas the sea-wall is a boundary to a feeding ground beyond which the bird cannot go. Better still, the inland bird has an immense advantage over the other in travelling to and from his nest in bad weather. When the wind blows strong from the sea the seaside bird must perpetually fight against it and win his home by sheer muscular exertion.

The other bird, able to go foraging to this side or that, according to the way the wind blows, can always have the wind as a help instead of a hindrance.

Somerset also possesses a long coast-line and some miles of sea-cliffs, but the colonies of jackdaws found here are small compared with those of the Mendip range. The inland-cliff breeding daws that inhabit the valley of the Somerset Axe alone probably greatly outnumber all the daws in Middlesex, or Surrey, or Essex.

Finally, besides the cliffs and woods, there are the old towns and villages--small towns and villages with churches that are almost like cathedrals. No county in England is richer in noble churches, and no kind of building seems more attractive to the "ecclesiastical daw" than the great Perpendicular tower of the Glastonbury type, which is so common here.

Of the old towns which the bird loves and inhabits in numbers, Wells comes first. If Wells had no birds it would still be a city one could not but delight in. There are not more than half a dozen towns in all the country where (if I were compelled to live in towns) life would not seem something of a burden; and of these, two are in Somerset--Bath and Wells.

Of the former something will be said further on: Wells has the first place in my affections, and is the one town in England the sight of which in April and early May, from a neighbouring hill, has caused me to sigh with pleasure. Its cathedral is assuredly the loveliest work of man in this land, supremely beautiful, even without the multitude of daws that make it their house, and may be seen every day in scores, looking like black doves perched on the stony heads and hands and shoulders of that great company of angels and saints, apostles, kings, queens, and bishops, that decorate the wonderful west front.

For in this building--not viewed as in a photograph or picture, nor through the eye of the mere architect or archaeologist, who sees the gem but not the setting--nature and man appear to have worked together more harmoniously than in others.

But it is hard to imagine a birdless Wells. The hills, beautiful with trees and grass and flowers, come down to it; cattle graze on their slopes; the peewit has its nest in their stony places, and the kestrel with quick-beating wings hangs motionless overhead. Nature is round it, breathing upon and touching it caressingly on every side; flowing through it like the waters that gave it its name in olden days, that still gush with noise and foam from the everlasting rock, to send their crystal currents along the streets. And with nature, in and around the rustic village-like city, live the birds. The green woodpecker laughs aloud from the group of old cedars and pines, hard by the cathedral close--you will not hear that woodland sound in any other city in the kingdom; and the rooks caw all day from the rookery in the old elms that grow at the side of the palace moat. But the cathedral daws, on account of their numbers, are the most important of the feathered inhabitants of Wells. These city birds are familiarly called "Bishop's Jacks," to distinguish them from the "Ebor Jacks," the daws that in large numbers have their home and breeding-place in the neighbouring cliffs, called the Ebor Rocks.

The Ebor daws are but the first of a succession of colonies extending along the side of the Cheddar valley. A curious belief exists among the people of Wells and the district, that the Ebor Jacks make better pets than the Bishop's Jacks. If you want a young bird you have to pay more for one from the rocks than from the cathedral. I was assured that the cliff bird makes a livelier, more intelligent and amusing pet than the other. A similar notion exists, or existed, at Hastings, where there was a saying among the fisher folks and other natives that "a Grainger daa is worth a ha'penny more than a castle daa." The Grainger rock, once a favourite breeding-place of the daws at that point, has long since fallen into the sea, and the saying has perhaps died out.

At Wells most of the cathedral birds--a hundred couples at least--breed in the cavities behind the stone statues, standing, each in its niche, in rows, tier above tier, on the west front. In April, when the daws are busiest at their nest-building, I have amused myself early every morning watching them flying to the front in a constant procession, every bird bringing his stick. This work is all done in the early morning, and about half-past eight o'clock a man comes with a barrow to gather up the fallen sticks--there is always a big barrowful, heaped high, of them; and if not thus removed the accumulated material would in a few days form a rampart or zareba, which would prevent access to the cathedral on that side.

It has often been observed that the daw, albeit so clever a bird, shows a curious deficiency of judg ment when building, in his persistent efforts to carry in sticks too big for the cavity. Here, for instance, each morning in turning over the litter of fallen material I picked up sticks measuring from four or five to seven feet in length. These very long sticks were so slender and dry that the bird was able to lift and to fly with them; therefore, to his corvine mind, they were suitable for his purpose. It comes to this: the daw knows a stick when he sees one, but the only way of testing its usefulness to him is to pick it up in his beak, then to try to fly with it.

If the stick is six feet long and the cavity will only admit one of not more than eighteen inches, he discovers his mistake only on getting home. The

question arises: Does he continue all his life long repeating this egregious blunder? One can hardly believe that an old, experienced bird can go on from day to day and year to year wasting his energies in gathering and carrying building materials that will have to be thrown away in the end--that he is, in fact, mentally on a level with the great mass of meaner beings who forget nothing and learn nothing.

It is not to be doubted that the daw was once a builder in trees, like all his relations, with the exception of the cliff-breeding chough. He is even capable of reverting to the original habit, as I know from an instance which has quite recently come to my knowledge. In this case a small colony of daws have been noticed for several years past breeding in stick nests placed among the clustering foliage of a group of Scotch firs. This colony may have sprung from a bird hatched and reared in the nest of a carrion crow or magpie. Still, the habit of breeding in holes must be very ancient, and considering that the jackdaw is one of the most intelligent of our birds, one cannot but be astonished at the rude, primitive, blundering way in which the nest-building work is generally performed.

The most we can see by carefully watching a number of birds at work is that there appears to be some difference with regard to intelligence between bird and bird. Some individuals blunder less than others; it is possible that these have learned something from experience; but if that be so, their better way is theirs only, and their young will not inherit it.

One morning at Wells as I stood on the cathedral green watching the birds at their work, I witnessed a rare and curious scene--one amazing to an ornithologist.

A bird dropped a stick--an incident that occurred a dozen times or oftener any minute at that busy time; but in this instance the bird had no sooner let the stick fall than he rushed down after it to attempt its recovery, just as one may see a sparrow drop a feather or straw, and then dart down after it and often recover it before it touches the ground. The heavy stick fell straight and fast on to the pile of sticks already lying on the pavement, and instantly the daw was down and had it in his beak, and thereupon laboriously flew up to his nesting-place, which was forty to fifty feet high.

At the moment that he rushed down after the falling stick two other daws that happened to be standing on ledges above dropped down after him, and copied his action by each picking up a stick and flying with it to their nests. Other daws followed suit, and in a few minutes there was a stream of descending and ascending daws at that spot, every ascending bird with a stick in his beak. It was curious to see that although sticks were lying in hundreds on the pavement along the entire breadth of the west front, the daws continued coming down only at that spot where the first bird had picked up the stick he had dropped. By and by, to my regret, the birds suddenly took alarm at something and rose up, and from that moment not one descended.

Presently the man came round with his rake and broom and barrow to tidy up the place. Before beginning his work he solemnly made the following remark: "Is it not curious, sir, considering the distance the birds go to get their sticks, and the work of carrying them, that they never, by any chance, think to come down and pick up what they have dropped!" I replied that I had heard the same thing said before, and that it was in all the books; and then I told him of the scene I had just witnessed.

He was very much surprised, and said that such a thing had never been witnessed before at that place.

It had a disturbing effect on him, and he appeared to me to resent this departure from their old ancient conservative ways on the part of the cathedral birds.

For many mornings after I continued to watch the daws until the nest-building was finished, without witnessing any fresh outbreak of intelligence in the colony: they had once more shaken down into the old inconvenient traditional groove, to the manifest relief of the man with the broom and barrow.

Bath, like Wells, is a city that has a considerable amount of nature in its composition, and is set down in a country of hills, woods, rocks and streams, and is therefore, like the other, a city loved by daws and by many other wild birds. It is a town built of white stone in the hollow of an oblong basin, with the river Avon flowing through it; and though perhaps too large for perfect beauty, it is exceedingly pleasant. Its "stone walls do not a prison make,"

since they do not shut you out from rural sights and sounds: walking in almost any street, even in the lowest part, in the busiest, noisiest centre of the town, you have but to lift your eyes to see a green hill not far away; and viewed from the top of one of these hills that encircle it, Bath, in certain favourable states of the atmosphere, wears a beautiful

look. One afternoon, a couple of miles out, I was on the top of Barrow Hill in a sudden, violent storm of rain and wind; when the rain ceased, the sun burst out behind me, and the town, rain-wet and sun-flushed, shone white as a city built of whitest marble against the green hills and black cloud on the farther side. Then on the slaty blackness appeared a complete and most brilliant rainbow, on one side streaming athwart the green hill and resting on the centre of the town, so that the high, old, richly-decorated Abbey Church was seen through a band of green and violet mist. That storm and that rainbow, seen by chance, gave a peculiar grace and glory to Bath, and the bright, unfading picture it left in memory has perhaps become too much associated in my mind with the thought of Bath, and has given me an exaggerated idea of its charm.

When staying in Bath in the winter of 1898-9 I saw a good deal of bird life even in the heart of the town. At the back of the house I lodged in, in New King Street, within four minutes' walk of the Pump Room, there was a strip of ground called a garden, but with no plants except a few dead stalks and stumps and two small leafless trees. Clothes-lines were hung there, and the ground was littered with old bricks and rubbish, and at the far end of the strip there was a fowl-house with fowls in it, a small shed, and a wood-pile. Yet to this unpromising-looking spot came a considerable variety of birds. Starlings, sparrows, and chaffinches were the most numerous, while the blackbird, thrush, robin, hedge-sparrow and wren were each represented by a pair. The wrens lived in the wood-pile, and were the only members of the little feathered community that did not join the others at table when crumbs and scraps were thrown out.

It was surprising to find all or most of these birds evidently wintering on that small plot of ground in the middle of the town, solely for the sake of the warmth and shelter it afforded them, and the chance crumbs that came in their way. It is true that I fed them regularly, but they were all there before I came.

Yet it was not an absolutely safe place for them, being much infested by cats, especially by a big black one who was always on the prowl, and who had a peculiarly murderous gleam in his luminous yellow orbs when he crouched down to watch or attempted to stalk them. One could not but

imagine that the very sight of such eyes in that black, devilish face would have been enough to freeze their blood with sudden terror, and make them powerless to fly from him. But it was not so: he could neither fascinate nor take them by surprise. No sooner would he begin to practise his wiles than all the population would be up in arms--the loud, sharp summons of the blackbird sounding first; then the starlings would chatter angrily, the thrush scream, the chaffinches begin to *pink-pink* with all their might, and the others would join in, even the small hideling wrens coming out of their fortress of faggots to take part in the demonstration. Then puss would give it up and go away, or coil himself up and go to sleep on the sloping roof of the tiny shed or in some other sheltered spot; peace and quiet would once more settle on the little republic, and the birds would be content to dwell with their enemy in their midst in full sight of them, so long as he slept or did not watch them too narrowly.

Finding that blue tits were among the visitors at the back, I hung up some lumps of suet and a cocoa-nut to the twigs of the bushes. The suet was immediately attacked, but judging from the suspicious way in which they regarded the round brown object swinging in the wind, the Bath tits had never before been treated to a cocoa-nut.

But though suspicious, it was plain that the singular object greatly excited their curiosity. On the second day they made the discovery that it was a new and delightful dish invented for the benefit of the blue tits, and from that time they were at it at all hours, coming and going from morning till night. There were six of them, and occasionally they were all there at once, each one anxious to secure a place, and never able when he got one to keep it longer than three or four seconds at a time.

Looking upon them from an upper window, as they perched against and flitted round and round the suspended cocoa-nut, they looked like a gathering of very large pale-blue flies flitting round and feeding on medlar.

No doubt the sparrow is the most abundant

species in Bath--I have got into a habit of not noticing that bird, and it is as if I did not see him; but after him the starling is undoubtedly the most numerous. He is, we know, increasing everywhere, but in no other town in England have I

found him in such numbers. He is seen in flocks of a dozen to half a hundred, busily searching for grubs on every lawn and green place in and round the town, and if you go up to some elevated spot so as to look down upon Bath, you will see flocks of starlings arriving and departing at all points.

As you walk the streets their metallic *clink-clink-clink* sounds from all quarters--small noises which

to most men are lost among the louder noises of a populous town. It is as if every house had a peal of minute bells hidden beneath the tiles or slates of the roof, or among the chimney-pots, that they were constantly being rung, and that every bell was cracked.

The ordinary or unobservant person sees and

hears far more of the jackdaw than of any other bird in Bath. Daws are seen and heard all over the town, but are most common about the Abbey, where they soar and gambol and quarrel all day long, and when they think that nobody is looking, drop down to the streets to snatch up and carry off any eatable-looking object that catches their eye.

It was here at this central spot, while I stood one day idly watching the birds disporting themselves about the Abbey and listened to their clamour, that certain words of Ruskin came into my mind, and I began to think of them not merely with admiration, as when I first read them long ago, but critically.

Ruskin, one of our greatest prose writers, is usually at his best in the transposition of pictures into words, his descriptions of what he has seen, in nature and art, being the most perfect examples of "word painting" in the language. Here his writing is that of one whose vision is not merely, as in the majority of men, the most important and intellectual of the senses, but so infinitely more important than all the others, and developed and trained to so extraordinary a degree, as to make him appear like a person of a single sense. We may say that this predominant sense has caused, or fed upon, the decay of the others. This is to me a defect in the author I most admire; for though he makes me see, and delight in seeing, that which was previously hidden, and all things gain in beauty and splendour, I yet miss something from the picture, just as I should miss light and colour from a description of nature, however beautifully written, by a man whose sense of sight was nothing or next to nothing to him, but whose other senses were all developed to the highest state of perfection.

No doubt Ruskin is, before everything, an artist: in other words, he looks at nature and all visible things with a purpose, which I am happily without: and the reflex effect of his purpose is to make nature to him what it can never appear to me--a painted canvas. But this subject, which I have touched on in a single sentence, demands a volume.

Ruskin wrote of the cathedral daws, "That drift of eddying black points, now closing, now scattering, now settling suddenly into invisible places

among the bosses and flowers, the crowd of restless birds that fill the whole square with that strange clangour of theirs, so harsh and yet so soothing."

For it seemed to me that he had seen the birds but had not properly heard them; or else that to his mind the sound they made was of such small consequence in the effect of the whole scene--so insignificant an element compared with the sight

of them--that it was really not worth attending to and describing accurately.

Possibly, in this particular case, when in speaking of the daws he finished his description by throwing in a few words about their voices, he was thinking less of the impression on his own mind, presumably always vague about natural sounds, than of what the poet Cowper had said in the best passage in his best work about "sounds harsh and inharmonious in themselves," which are yet able to

produce a soothing effect on us on account of the peaceful scenes amid which they are heard.

Cowper's notion of the daw's voice, by the way, was just as false as that expressed by Ruskin, as we may find in his paraphrase of Vincent Bourne's lines to that bird:--

There is a bird that by his coat,
And by the hoarseness of his note
Might be supposed a crow.

Now the daw is capable at times of emitting

both hoarse and harsh notes, and the same may perhaps be said of a majority of birds; but his usual note--the cry or caw varied and inflected a hundred ways, which we hear every day and all day long where daws abound--is neither harsh like the crow's, nor hoarse like the rook's. It is, in fact, as unlike the harsh, grating caw of the former species as the clarion call of the cock is unlike the grunting of swine. It may not be described as bell-like nor metallic, but it is loud and clear, with an engaging wildness in it, and, like metallic sounds, far-reaching; and of so good a quality that very little more would make it ring musically.

Sometimes when I go into this ancient abbey church, or into some cathedral, and seating myself, and looking over a forest of bonnets, see a pale young curate with a black moustache, arrayed in white vestments, standing before the reading-desk, and hear him gabbling some part of the Service in a continuous buzz and rumble that roams like a gigantic blue-bottle through the vast dim interior, then I, not following him--for I do not know where he is, and cannot find out however much I should like to--am apt to remember the daws out of doors, and to think that it would be well if that young man would but climb up into the highest tower, or on to the roof, and dwell there for the space of a year listening to them; and that he would fill his mouth with polished pebbles, and medals, and coins and seals and seal-rings, and small porcelain cats and dogs, and little silver pigs, and other objects from the chatelaines of his lady admirers, and strive to imitate that clear, penetrating sound of the bird's voice, until he had mastered the rare and beautiful arts of voice production and distinct understandable speech.

To go back to Cowper--the poet who has been much in men's thoughts of late, and who appears to us as perhaps the most modern-minded of those who ceased to live a century ago. Undoubtedly he was as bad a naturalist as any singer before or after him, and as any true poet has a perfect right to be. As bad, let us say, as Shakespeare and Wordsworth and Tennyson. He does not, it is true, confound the sparrow and hedge-sparrow like Wordsworth, nor confound the white owl with the brown owl like Tennyson, nor puzzle the ornithologist with a "sea-blue bird of March." But we must not forget that he addressed some verses to a nightingale heard on New Year's Day. It is clear that he did not know the crows well, for in a letter of May 10, 1780, to his friend Newton, he writes: "A crow, rook, or raven, has built a nest in one of the young elm-trees, at the side of Mrs Aspray's orchard." But when he wrote those words--

Sounds inharmonious in themselves, and harsh,
Yet heard in scenes where peace for ever reigns,
And only there, please highly for their sake--

words which I have suggested misled Ruskin, and have certainly misled others--he, Cowper, knew better. His real feeling, and better and wiser thought, is expressed in one of his incomparable letters (Hayley, vol. ii. p. 230)--

"My green-house is never so pleasant as when we are just on the point of surrendering it.... I sit with all the windows and the door wide open, and am regaled with the scent of every flower in a garden as full of flowers as I have known how to make it. We keep no bees, but if I lived in a hive I could hardly have more of their music. All the bees in the neighbourhood resort to a bed of mignonette opposite to the window, and pay me for the honey they get out of it by a hum, which, though rather monotonous, is as agreeable to my ears as the whistling of my linnets. All the sounds that nature utters are delightful, at least in this country. I should not perhaps find the roaring of lions in Africa, or of bears in Russia, very pleasing; but I know no beast in England whose voice

I do not account as musical, save and except always the braying of an ass. The notes of all our birds and fowls please me, without one exception. I should not indeed think of keeping a goose in a cage that I might hang him up in the parlour for the sake of his melody, but a goose upon a common, or in a farmyard, is no bad performer; and as to insects, if the black beetle, and beetles indeed of all hues, will keep out of my way, I have no objection to any of the rest; on the contrary, in whatever key they sing, from the gnat's fine treble to the bass of the bumble-bee, I admire all. Seriously, however, it strikes me as a very observable instance of providential kindness to men, that such an exact accord has been contrived between his ear and the sounds with which, at least in a rural situation, it is almost every moment visited."

Who has not felt the truth of this saying, that all natural sounds heard in their proper surroundings are pleasing; that even those which we call

harsh do not distress, jarring or grating on our nerves, like artificial noises! The braying of the donkey was to Cowper the one exception in animal life; but he never heard it in its proper conditions.

I have often listened to it, and have been deeply impressed, in a wild, silent country, in a place where herds of semi-wild asses roamed over the plains; and the sound at a distance had a wild expression that accorded with the scene, and owing to its much greater power effected the mind more than the trumpeting of wild swans, and shrill neighing of wild horses, and other far-reaching cries of wild animals.

About the sounds emitted by geese in a state of nature, and the effect produced on the mind, I shall have something to say in a chapter on that bird.

CHAPTER IV

EARLY SPRING IN SAVERNAKE FOREST

When the spring-feeling is in the blood, infecting us with vague longings for we know not what; when we are restless and seem to be waiting for some obstruction to be removed--blown away by winds, or washed away by rains--some change

that will open the way to liberty and happiness,--the feeling not unfrequently takes a more or

less definite form: we want to go away somewhere, to be at a distance from our fellow-beings, and nearer, if not to the sun, at all events to wild nature.

At such times I think of all the places where I should like to be, and one is Savernake; and thither in two following seasons I have gone to ramble day after day, forgetting the world and myself in its endless woods.

It is not that spring is early there; on the contrary, it is actually later by many days than in the surrounding country. It is flowerless at a time when, outside the forest, on southern banks and by the hedge-side, in coppices and all sheltered spots, the firstlings of the year are seen--purple and white and yellow. The woods, which are

composed almost entirely of beech and oak, are leafless. The aspect on a dull cold day is somewhat cheerless. On the other hand, there is that

largeness and wildness which accord with the spring mood; and there are signs of the coming change even in the greyest weather. Standing in some wide green drive or other open space, you see all about you acres on acres, miles on miles, of majestic beeches, and their upper branches and network of terminal twigs, that look at a distance like heavy banked-up clouds, are dusky red and purple with the renewed life that is surging in them. There are jubilant cries of wild creatures that have felt the seasonal change far more keenly than we are able to feel it. Above everything, we find here that solitariness and absence of human interest now so rare in England. For albeit social creatures in the main, we are yet all of us at times hermits in heart, if not exactly wild men of the woods; and that solitude which we create by shutting ourselves from the world in a room or a house, is but a poor substitute--nay, a sham: it is to immure ourselves in a cage, a prison, which hardly serves to keep out the all-pervading atmosphere of miserable conventions, and cannot refresh and invigorate us. There are seasons and moods when even the New Forest does not seem sufficiently remote from life: in its most secluded places one is always liable to encounter a human being, an old resident, going about in the exercise of his commoner's rights; or else his ponies or cows or swine. These last, if they be not of some improved breed, may have a novel or quaint aspect, as of wild creatures, but the appearance is deceptive; as you pass they lift their long snouts from grubbing among the dead leaves to salute you with

a too familiar grunt--an assurance that William Rufus is dead, and all is well; that they are domestic, and will spend their last days in a stye,

and end their life respectably at the hands of the butcher.

At Savernake there is nothing so humanised as the pig, even of the old type; you may roam for long hours and see no man and no domestic animal.

You have heard that this domain is the property of some person, but it seems like a fiction. The forest is nature's and yours. There you are at liberty to ramble all day unchallenged by any one; to walk, and run to warm yourself; to disturb a herd of red deer, or of fallow deer, which are more numerous; to watch them standing still to gaze

back at you, then all with one impulse move rapidly away, showing their painted tails, keeping a kind of discipline, row behind row, moving over the turf with that airy tripping or mincing gait that strikes you as quaint and somewhat bird-like.

Or you may coil yourself up, adder-like, beside a thick hawthorn bush, or at the roots of a giant oak or beech, and enjoy the vernal warmth, while outside of your shelter the wind blows bleak and loud.

To lie or sit thus for an hour at a time listening to the wind is an experience worth going far to seek. It is very restorative. That is a mysterious voice which the forest has: it speaks to us, and somehow the life it expresses seems nearer, more intimate, than that of the sea. Doubtless because we are ourselves terrestrial and woodland in our origin; also because the sound is infinitely more varied as well as more human in character. There are sighings and moanings, and wails and shrieks, and wind-blown murmurings, like the distant confused talking of a vast multitude. A high wind

in an extensive wood always produces this effect of numbers. The sea-like sounds and rhythmic volleyings, when the gale is at its loudest, die away, and in the succeeding lull there are only low, mysterious agitated whisperings; but they are multitudinous; the suggestion is ever of a vast concourse--crowds and congregations, tumultuous or orderly,

but all swayed by one absorbing impulse, solemn or passionate. But not always moved simultaneously.

Through the near whisperings a deeper,

louder sound comes from a distance. It rumbles like thunder, falling and rising as it rolls onwards; it is antiphonal, but changes as it travels

nearer. Then there is no longer demand and response; the smitten trees are all bent one way,

and their innumerable voices are as one voice, expressing we know not what, but always something not wholly strange to us--lament, entreaty,

denunciation.

Listening, thinking of nothing, simply living in the sound of the wind, that strange feeling which is unrelated to anything that concerns us, of the life and intelligence inherent in nature, grows upon the mind. I have sometimes thought that never does the world seem more alive and watchful of us than on a still, moonlight night in a solitary wood, when the dusky green foliage is silvered by the beams, and all visible objects and the white lights and black shadows in the intervening spaces seem instinct with spirit. But it is not so. If the conditions be favourable, if we go to our solitude as the crystal-gazer to his crystal, with a

mind prepared, this faculty is capable of awaking and taking complete possession of us by day as well as by night.

As the trees are mostly beeches--miles upon

miles of great trees, many of them hollow-trunked from age and decay--the fallen leaves are an important element in the forest scenery. They lie

half a yard to a yard deep in all the deep hollows and dells and old water-worn channels, and where the ground is sheltered they cover acres of ground--millions and myriads of dead, fallen beech leaves.

These, too, always seem to be alive. It is a leaf that refuses to die wholly. When separated from the tree it has, if not immortality, at all events a second, longer life. Oak and ash and chestnut leaves fade from month to month and blacken, and finally rot and mingle with the earth, while the beech leaf keeps its sharp clean edges unbroken, its hard texture and fiery colour, its buoyancy and rustling incisive sound. Swept by the autumn winds into sheltered hollows and beaten down by rains, the leaves lie mingled in one dead, sodden mass for days and weeks at a time, and appear ready to mix with the soil; but frost and sun suck up the moisture and the dead come to life again.

They glow like fire, and tremble at every breath.

It was strange and beautiful to see them lying all around me, glowing copper and red and gold when the sun was strong on them, not dead, but sleeping like a bright-coloured serpent in the genial warmth; to see, when the wind found them, how they

trembled, and moved as if awakening; and as

the breath increased rose up in twos and threes and half-dozens here and there, chasing one another a little way, hissing and rustling; then all at once, struck by a violent gust, they would be up in thousands, eddying round and round in a dance, and, whirling aloft, scatter and float among the lofty branches to which they were once attached.

On a calm day, when there was no motion in

the sunlit yellow leaves below and the reddish-purple cloud of twigs above, the sounds of bird-life were the chief attraction of the forest. Of

these the cooing of the wood-pigeon gave me the most pleasure. Here some reader may remark

that this pigeon's song is a more agreeable sound than its plain cooing note. This, indeed, is perhaps thought little of. In most biographies of the bird it is not even mentioned that he possesses such a note. Nevertheless I prefer it to the song.

The song itself--the set melody composed of half a dozen inflected notes, repeated three or four times with little or no variation--is occasionally heard in the late winter and early spring, but at this time of the year it is often too husky or croaky to be agreeable. The songster has not yet thrown off his seasonal cold; the sound might sometimes proceed from a crow suffering from a catarrh. It improves as the season advances. The song is sometimes spelt in books:

Coo-coo-roo, coo-coo-roo.

A lady friend assures me the right words of this song are:

Take *two* cows, David.

She cannot, if she tries, make the bird say anything different, for these are the words she was taught to hear in the song, as a child, in Leicestershire.

Of course they are uttered with a great deal of emotion in the tone, David being tearfully, almost sobbingly, begged and implored to take two cows; the emphasis is very strong on the two--it is apparently a matter of the utmost consequence that David should not take one, nor three, nor any other number of cows, but just two.

In East Anglia I have been informed that what the bird really and truly says is--

My toe bleeds, Betty.

Many as are the species capable of articulate speech, as we may see by referring to any ornithological work, there is no bird in our woods whose notes more readily lend themselves to this childish fancy than the wood-pigeon, on account of the depth and singularly human quality of its voice. The song is a passionate complaint. One can fancy the human-like feathered creature in her green bower, pleading, upbraiding, lamenting; and, listening, we will find it easy enough to put it all into plain language: O swear not you love me, for you cannot be true,

O perjured wood-pigeon! Go from me--woo
Some other! Heart-broken I rue
That softness, ah me! when you cooed your false coo.
Soar to your new love--the creature in blue!
Who, who would have thought it of you!
And perhaps you consider her beau--
Oo--tiful! O you are too too cru--
Bid them come shoo--oot me, do, do!
Would I had given my heart to a hoo--
Oo-ting wood-owl, cuckoo, woodcock, hoopoo!

One morning, at a village in Berkshire, I was walking along the road, about twenty-five yards from a cottage, when I heard, as I imagined, the familiar song of the wood-pigeon; but it sounded too close, for the nearest trees were fifty yards distant. Glancing up at the open window of an upper room in the cottage, I made the discovery that my supposed pigeon was a four-year-old child who had recently been chastised by his mother and sent upstairs to do penance. There he sat by the open window, his face in his hands, crying, not as if his heart would break, but seeming to take a mournful pleasure in the rhythmical sound of his own sobs and moans; they had settled into a rising and falling *boo-hoo*, with regularly recurring long and short notes, agreeable to the ear, and very creditable to the little crier's musical capacity. The incident shows how much the pigeon's plaint resembles some human sounds.

The plain cooing note is so common in this order of birds that it may be regarded as the original and universal pigeon language, out of which the set songs have been developed, with, in most instances, but little change in the quality of the sound. In the multitude of species there are voices clear, resonant, thick, or husky, or guttural, hollow or booming, grating and grunting; but, however much they vary, you can generally detect the *pigeon* or *family* sound, which is more or less human-like. In some species the set song has almost superseded the plain single note, which has diminished to a mere murmur; in others, on the contrary, there is no song at all, unless the single unvarying *coo* can be called a song. In most species in the typical genus Columba the plain coo is quite distinct from the set song, but has at the same time developed into a kind of second song, the note being pleasantly modulated and repeated many times. We find this in the rock-dove: the curious guttural sounds composing its set song, which accompany the love antics of the male, are not musical, while the clear inflected cooing note is agreeable to most ears. It is a pleasing morning sound of the dove-cote; but the note, to be properly appreciated, must be heard in some dimly lighted ocean-cavern in which the bird breeds in its wild state.

The long-drawn, oft-repeated musical coo mingles with and is heard above the murmuring and lapping of the water beneath; the hollow chamber retains and prolongs the sound, and makes it more sonorous, and at the same time gives it something of mystery.

Of all the cooing notes of the different species I am acquainted with, that of the stock-dove, a pigeon with no set song, is undoubtedly the most attractive: next in order is that of the wood-pigeon on account of its depth and human-like character. And it is far from monotonous. In this wood in March I have often kept near a pigeon for half an hour at a time hearing it uttering its cooing note, repeated half a dozen or more times, at intervals of three or four minutes; and again and again the note has changed in length and power and modulation. In the profound stillness, on a windless day, of the vast beechen woods, these sonorous notes had a singularly beautiful effect.

After spending a short time in the forest, one might easily get the idea that it is a sanctuary for all the persecuted creatures of the crow family. It is not quite that; the ravens have been destroyed here as in most places; but the other birds of that tribe are so numerous that even the most bloodthirsty keeper might be appalled at the task of destroying them. The clearance would doubtless have been effected if this noble forest had passed, as so nearly happened, out of the hands of the family that have so long possessed it: that calamity was happily averted. Not only are the rooks there in legions, having their rookeries in the park, but, throughout the forest, daws, carrion crows, jays, and magpies are abundant. The jackdaws outnumber all the other species (rooks included) put together; they literally swarm, and their ringing, yelping cries may be heard at all hours of the day in any part of the forest. In March, when they are nesting, their numbers are concentrated in those parts of the wood where the trees, beech and oak, are very old and have hollow trunks. In some places you will find many acres of wood where every tree is hollow and apparently inhabited. Yet there are doubtless some hollow trees into which the daw is not permitted to intrude. The wood-owl is common here, and is presumably well able to hold his castle against all aggressors. If one could but climb into the airy tower, and, sitting invisible, watch the siege and defence and the many strange incidents of the war between these feathered foes! The daw, bold yet cautious, venturing a little way into the dim interior, with shrill threats of ejectment, ruffling his grey pate and peeping down with his small, malicious, serpent-like grey eyes; the owl puffing out his tiger-coloured plumage, and lifting to the light his pale, shield-like face and luminous eyes,--would indeed be a rare spectacle; and then, what hissings, snappings, and beak-clatterings, and shrill, cat-like, and yelping cries! But, although these singular contests go on so near us, a few yards above the surface, Savernake might be in the misty mid-region of Weir, or on the slopes of Mount Yanik, for all the chance we have of witnessing them.

An experience I had one day when I was new to the forest and used occasionally to lose myself, gave me some idea of the numbers of jackdaws breeding in Savernake. During my walk I came to a spot where all round me and as far as could be seen the trees were in an advanced state of decay: not only were they hollow and rotten within, but the immense horizontal branches and portions of the trunks were covered with a thick crop of fern, which, mixed with dead grass and moss, gave the dying giants of the forest a strange, ragged and desolate appearance. Many a time looking at one of these trees I have been reminded of Holman Hunt's forlorn Scapegoat. Here the daws had their most populous settlement. As I advanced, the dead twigs and leaves crackling beneath my feet, they rose up everywhere, singly and in twos and threes and half-dozens, darting hurriedly away and disappearing among the trees before me. The alarm-note they emit at such times is like their usual yelping call subdued to a short, querulous chirp; and this note now sounded before me and on either hand, at a distance of about one hundred yards, uttered continually by so many birds that their voices mingled into a curious sharp murmur. Tired of walking, I sat down on a root in the shelter of a large oak, and remained there perfectly motionless for about an hour. But the birds never lost their suspicion; all the time the distant subdued tempest of sharp notes went on, occasionally dying down until it nearly ceased, then suddenly rising and spreading again until I was ringed round with the sound. At length the loud, sharp invitation or order to fly was given and taken up by many birds; then, through the opening among the trees before me, I saw them rise in a dense flock and circle about at a distance: other flocks rose on the right and left hands and joined the first; and finally the whole mass come slowly overhead as if to explore; but when the foremost birds were directly over me the flock divided into two columns, which deployed to the right and left, and at a distance poured again into the trees.

There could not have been fewer than two thousand birds in the flock that came over me, and they were probably all building in that part of the forest.

The daw, whether tame or distrustful of man, is always interesting. Here I was even more interested in the jays, and it was indeed chiefly for the pleasure of seeing them, when they are best to look at, that I visited this forest. I had also formed the idea that there was no place in England where the jay could be seen to better advantage, as they are, or until recently were, exceedingly abundant at Savernake, and were not in constant fear of the keeper and his everlasting gun. Here one could witness their early spring assemblies, when the jay, beautiful at all times, is seen at his very best.

It is necessary to say here that this habit of the jay does not appear to be too well known to our ornithologists. When I stated in a small work on *British Birds* a few years ago that jays had the custom of congregating in spring, a distinguished naturalist, who reviewed the book in one of the papers, rebuked me for so absurd a statement, and informed me that the jay is a solitary bird except at the end of summer and in the early autumn, when they are sometimes seen in families. If I had not made it a rule never to reply to a critic, I could have informed this one that I knew exactly where his knowledge of the habits of the jay was derived-that it dated back to a book published ninety-nine years ago. It was a very good book, and all it contains, some errors included, have been incorporated in most of the important ornithological works which have appeared during the

nineteenth century. But though my critic thus

"wrote it all by rote," according to the books, "he did not write it right." The ancient error has not, however, been repeated by all writers on the subject. Seebohm, in his *History of British Birds*, wrote: "Sometimes, especially in Spring, fortune may favour you, and you will see a regular gathering of these noisy birds.... It is only at this time that the jay displays a social disposition; and the birds may often be heard to utter a great variety of notes, some of the modulations approaching almost to a song."

The truth of the statement I have made that

most of our writers on birds have strictly followed Montague in his account of the jay's habits, unmistakably shows itself in all they say about the

bird's language. Montagu wrote in his famous *Dictionary of Birds* (1802):--

"Its common notes are various, but harsh;

will sometimes in spring utter a sort of song in a soft and pleasing manner, but so low as not to be heard at any distance; and at intervals introduce the bleatings of a Lamb, mewing of a Cat, the note of a Kite or Buzzard, hooting of an Owl, and even the neighing of a Horse.

"These imitations are so exact, even in a natural wild state, that we have frequently been deceived."

This description somewhat amplified, and the

wording varied to suit the writer's style, has been copied into most books on British birds--the lamb and the cat, and the kite and the horse, faithfully appearing in most cases. Yet it is certain that if all the writers had listened to the jay's vocal performances for themselves, they would have given a

different account. It is not that Montagu was wrong: he went to nature for his facts and put down what he heard, or thought he heard, but the particular sounds which he describes they would not have heard.

My experience is, that the same notes and phrases are not ordinarily heard in any two localities; that the bird is able to emit a great variety of sounds--some highly musical; that he is also a great mimic in a wild irregular way, mixing borrowed notes with his own, and flinging them out anyhow, so that there is no order nor harmony, and they do not form a song.

But he also has a real song, which may be heard in any assembly of jays and from some male birds after the congregating season is over and breeding is in progress. This singing of the jay is somewhat of a puzzle, as it is not the same song in any two places, and gives one the idea that there is no inherited and no traditional song in this species, but that each bird that has a song has invented it for himself. It varies from "a sort of low song,"

as Montagu said,--a soft chatter and warble which one can just hear at a distance of thirty or forty yards,--to a song composed of several musical notes harmoniously arranged, which may be heard distinctly a quarter of a mile away. This set and far-reaching song is rare, but some birds have a single very powerful and musical note, or short phrase, which they repeat at regular intervals by way of song. If by following up the sound one can get near enough to the tree where the meeting is being held to see what is going on, it is most interesting to watch the vocalist, who is like a leader, and who, perched quietly, continues to repeat that one powerful, unchanging, measured sound in the midst of a continuous concert of more or less musical sounds from the other birds.

What I should very much like to know is, whether these powerful and peculiar notes, phrases, and songs of the jay, which are clearly not imitations of other species, are repeated year after year by the birds in the same localities, or are dropped for ever or forgotten at the end of each season. It is hard for me to find this out, because I do not as a rule revisit the same places in spring, and on going to a new or a different spot I find that the birds utter different sounds. Again, the places where jays assemble in numbers are very few and far between. It is true, as an observant gamekeeper once said to me, that if there are as many as half a dozen to a dozen jays in any wood they will contrive to hold a meeting; but when the birds are few and much persecuted, it is difficult to see and hear them at such times, and when seen and heard, no adequate idea is formed of the beauty of their displays, and the power and variety of their language, as witnessed in localities where they are numerous, and fear of the keeper's gun has not damped their mad, jubilant spirits.

In genial weather the jays' assembly may be held at any hour, but is most frequently seen during the early part of the day: on a fine warm morning in March and April one can always count on witnessing an assembly, or at all events of hearing the birds, in any wood where they are fairly common and not very shy. They are so vociferous and so conspicuous to the eye during these social intervals, and at the same time so carried away by excitement, that it is not only easy to find and see them, but possible at times to observe them very closely.

The loud rasping alarm-and angry-cry of the jay is a sound familiar to every one; the cry used by the bird to call his fellows together is somewhat different. It resembles the cry or call of the carrion crow, in localities where that bird is not persecuted, when, in the love season, he takes his stand on the top of the nesting-tree and calls with a prolonged, harsh, grating, and exceedingly powerful note, many times repeated. The jay's call has the same grating or grinding character, but is louder, sharper, more prolonged, and in a quiet atmosphere may be heard distinctly a mile away. The wood is in an uproar when the birds assemble and scream in concert while madly pursuing one another over the tall trees.

At such times the peculiar flight of the jay is best seen and is very beautiful. In almost all birds that have short, round wings, as we may see in our little wren, and in game birds, and the sparrow-hawk, and several others, the wing-beats are exceedingly rapid. This is the case with the magpie; the quickness of the wing-beats causes the black and white on the quills to mingle and appear a misty grey; but at short intervals the bird glides and the wings appear black and white again. The jay, although his wings are so short and round, when not in a hurry progresses by means of comparatively slow, measured wing-beats, and looks as if swimming rather than flying.

It is when the gathered birds all finally settle on a tree that they are most to be admired. They will sometimes remain on the spot for half an hour or longer, displaying their graces and emitting the extraordinary medley of noises mixed with musical sounds. But they do not often sit still at such times; if there are many birds, and the excitement is great, some of them are perpetually moving, jumping and flitting from branch to branch, and springing into the air to wheel round or pass over the tree, all apparently intent on showing off their various colours--vinaceous brown, sky blue, velvet black, and glistening white--to the best advantage.

Again and again, when watching these gatherings at Savernake and at other places where jays

abound, I have been reminded of the description given by Alfred Russel Wallace of the bird of paradise assemblies in the Malayan region. Our jay in some ways resembles his glorious Eastern relation; and although his lustre is so much less, he is at his very best not altogether unworthy of being called the British Bird of Paradise.

CHAPTER V

A WOOD WREN AT WELLS

East of Wells Cathedral, close to the moat surrounding the bishop's palace, there is a beautifully

wooded spot, a steep slope, where the birds had their headquarters. There was much to attract them there: sheltered by the hill behind, it was a warm corner, a wooded angle, protected by high old stone walls, dear to the redstart, masses of ivy, and thickets of evergreens; while outside the walls were green meadows and running water.

When going out for a walk I always passed through this wood, lingering a little in it; and when I wanted to smoke a pipe, or have a lazy hour to myself among the trees, or sitting in the sun, I almost invariably made for this favourite spot.

At different hours of the day I was a visitor, and there I heard the first spring migrants on their arrival--chiff-chaff, willow wren, cuckoo, redstart, blackcap, white-throat. Then, when April was drawing to an end, I said, There are no more to come. For the wryneck, lesser white-throat, and garden warbler had failed to appear, and the few nightingales that visit the neighbourhood had settled down in a more secluded spot a couple of miles away, where the million leaves in coppice and brake were not set a-tremble by the melodious thunder of the cathedral chimes.

Nevertheless, there was another still to come, the one I perhaps love best of all. On the last day of April I heard the song of the wood wren, and at once all the other notes ceased for a while to interest me. Even the last comer, the mellow blackcap, might have been singing at that spot since February, like the wren and hedge-sparrow, so familiar and workaday a strain did it seem to have compared with this late warbler. I was

more than glad to welcome him to that particular spot, where if he chose to stay I should have him so near me.

It is well known that the wood wren can only be properly seen immediately after his arrival in this country, at the end of April or early in May, when the young foliage does not so completely hide his slight unresting form, as is the case afterwards.

For he, too, is green in colour; like Wordsworth's green linnet,

A brother of the leaves he seems.

There is another reason why he can be seen so much better during the first days of his sojourn with us: he does not then keep to the higher parts of the tall trees he frequents, as his habit is later, when the air is warm and the minute winged insects on which he feeds are abundant on the upper sun-touched foliage of the high oaks and beeches. On account of that ambitious habit of the wood wren there is no bird with us so difficult to observe; you may spend hours at a spot, where his voice sounds from the trees at intervals of half a minute to a minute, without once getting a glimpse of his form. At the end of April the trees are still very thinly clad; the upper foliage is but an airy garment, a slight golden-green mist, through which the sun shines, lighting up the dim interior, and making the bed of old fallen beech-leaves look like a floor of red gold. The small-winged insects, sun-loving and sensitive to cold, then hold their revels near the surface; and the bird, too, prefers the neighbourhood of the earth. It was so in the case of the wood wren I observed at Wells, watching him on several consecutive days, sometimes for an hour or two at a stretch, and generally more than once a day. The spot where he was always to be found was quite free from underwood, and the trees were straight and tall, most of them with slender, smooth boles. Standing there, my figure must have looked very conspicuous to all the small birds in the place; but for a time it seemed to me that the wood wren paid not the slightest attention to my presence; that as he wandered hither and thither in sunlight and shade at his own sweet will, my motionless form was no more to him than a moss-grown stump or grey upright stone. By and by it became apparent that the bird knew me to be no stump or stone, but a strange living creature whose appearance greatly interested him; for invariably, soon after I had taken up my position, his careless little flights from twig to twig and from tree to tree brought him nearer, and then nearer, and finally near me he would remain for most of the time. Sometimes he would wander for a distance of forty or fifty yards away, but before long he would wander back and be with me once more, often perching so near that the most delicate shadings of his plumage were as distinctly seen as if I had had him perched on my hand.

The human form seen in an unaccustomed place always excites a good deal of attention among the birds; it awakes their curiosity, suspicion, and alarm. The wood wren was probably curious and nothing more; his keeping near me looked strange only because he at the same time appeared so wholly absorbed in his own music. Two or three times I tried the experiment of walking to a distance of fifty or sixty yards and taking up a new position; but always after a while he would drift thither, and I would have him near me, singing and moving, as before.

I was glad of this inquisitiveness, if that was the bird's motive (that I had unconsciously fascinated him I could not believe); for of all the wood wrens I have seen this seemed the most beautiful, most graceful in his motions, and untiring in song. Doubtless this was because I saw him so closely, and for such long intervals. His fresh yellowish-green upper and white under plumage gave him a wonderfully delicate appearance, and these colours harmonised with the tender greens of the opening leaves and the pale greys and silvery whites of the slender boles.

Seebohm says of this species: "They arrive in our woods in marvellously perfect plumage.

In the early morning sun they look almost as delicate a yellowish-green as the half-grown leaves amongst which they disport themselves. In the hand the delicate shading of the eye-stripe, and the margin of the feathers of the wings and tail, is exquisitely beautiful, but is almost all lost under the rude handling of the bird-skinner."

The concluding words sound almost strange; but it is a fact that this sylph-like creature is sometimes shattered with shot and its poor remains operated on by the bird-stuffer. Its beauty "in the hand" cannot compare with that exhibited when it lives and moves and sings. Its appearance during flight differs from that of other warblers on account of the greater length and sharpness of the wings. Most warblers fly and sing hurriedly; the wood wren's motions, like its song, are slower, more leisurely, and more beautiful. When moved by the singing passion it is seldom still for more than a few moments at a time, but is continually passing from branch to branch, from tree to tree, finding a fresh perch from which to deliver its song on each occasion. At such times it has the appearance of a delicately coloured miniature kestrel or hobby. Most lovely is its appearance when it begins to sing in the air, for then the long sharp wings beat time to the first clear measured notes, the prelude to the song. As a rule, however, the flight is silent, and the song begins when the new perch is reached--first the distinct notes that are like musical strokes, and fall faster and faster until they run and swell into a long passionate trill--the woodland sound which is like no other.

Charming a creature as the wood wren appears when thus viewed closely in the early spring-time, he is not my favourite among small birds because of his beauty of shape and colour and graceful motions, which are seen only for a short time, but on account of his song, which lasts until September; though I may not find it very easy to give a reason for the preference.

It comforts me a little in this inquiry to remember that Wordsworth preferred the stock-dove to the nightingale--that "creature of ebullient heart." The poet was a little shaky in his ornithology at times; but if we take it that he meant the ring-dove, his preference might still seem strange to some. Perhaps it is not so very strange after all.

If we take any one of the various qualities which we have agreed to consider highest in bird-music, we find that the wood wren compares badly with his fellow-vocalists--that, measured by this standard, he is a very inferior singer. Thus, in variety, he cannot compare with the thrush, garden-warbler, sedge-warbler, and others; in brilliance and purity of sound with the nightingale, blackcap, etc.; in strength and joyousness with the skylark; in mellowness with the blackbird; in sprightliness with the goldfinch and chaffinch; in sweetness with the wood-lark, tree-pipit, reed-warbler, the chats and wagtails, and so on to the end of all the qualities which we regard as important. What, then, is the charm of the wood wren's song? The sound is unlike any other, but that is nothing, since the same can be said of the wryneck and cuckoo and grasshopper warbler. To many persons the wood wren's note is a bird-sound and nothing more, and it may even surprise them to hear it called a song. Indeed, some ornithologists have said that it is not a song, but a call or cry, and it has also been described as "harsh."

I here recall a lady who sat next to me on the coach that took me from Minehead to Lynton.

The lady resided at Lynton, and finding that I was visiting the place for the first time, she proceeded to describe its attractions with fluent enthusiasm. When we arrived at the town, and were moving very slowly into it, my companion turned and examined my face, waiting to hear the expressions of rapturous admiration that would fall from my lips. Said I, "There is one thing you can boast of in Lynton. So far as I know, it is the only town in the country where, sitting in your own room with the windows open, you can listen to the song of the wood wren." Her face fell. She had never heard of the wood wren, and when I pointed to the tree from which the sound came and she listened and heard, she turned away, evidently too disgusted to say anything. She had been wasting her eloquence on an unworthy subject--one who was without appreciation for the sublime and beautiful in nature. The wild romantic Lynn, tumbling with noise and foam over its rough stony bed, the vast wooded hills, the piled-up black rocks (covered in places with beautiful red and blue lettered advertisements), had been passed by in silence--nothing had stirred me but the chirping of a miserable little bird, which, for all that she knew or cared, might be a sparrow!

When we got down from the coach a couple of minutes later, she walked away without even saying good-bye.

There is no doubt that very many persons know and care as little about bird voices as this lady; but how about the others who do know and care a good deal--what do they think and feel about the song of the wood wren? I know two or three persons who are as fond of the bird as I am; and two or three recent writers on bird life have spoken of its song as if they loved it. The ornithologists have in most cases been satisfied to quote Gilbert White's description of Letter XIX.: "This last haunts only the tops of trees in high beechen woods, and makes a sibilous grasshopper-like noise now and then, at short intervals, shaking a little with its wings when it sings."

White was a little more appreciative in the case of the willow wren when he spoke of its "joyous, easy, laughing note"; yet the willow wren has had to wait a long time to be recognised as one of our best vocalists. Some years ago it was greatly praised by John Burroughs, who came over from America to hear the British songsters, his thoughts running chiefly on the nightingale, blackcap, throstle, and blackbird; and he was astonished to find that this unfamed warbler, about which the ornithologists had said little and the poets nothing, was one of the most delightful vocalists, and had a "delicious warble." He waxed indignant at our neglect of such a singer, and cried out that it had too fine a song to please the British ear; that a louder coarser voice was needed to come up to John Bull's standard of a good song.

No one who loves a hearty laugh can feel hurt at his manner of expressing himself, so characteristic of an American. Nevertheless, the fact remains that only since Burroughs' appreciation of the British song-birds first appeared, several years ago, the willow wren, which he found languishing in obscurity, has had many to praise it. At all events, the merits of its song are now much more freely acknowledged than they were formerly.

Perhaps the wood wren's turn will come by and by. He is still an obscure bird, little known, or not known, to most people: we are more influenced by what the old writers have said than we know or like to believe; our preferences have mostly been made for us. The species which they praised and made famous have kept their places in popular esteem, while other species equally charming, which they did not know or said nothing about, are still but little regarded. It is hardly to be doubted that the wood wren would have been thought more of if Willughby, the Father of British Ornithology, had known it and expressed a high opinion of its song; or that it would have had millions to admire it if Chaucer or Shakespeare had singled it out for a few words of praise.

It is also probably the fact that those who are not students, or close observers of bird life, seldom know more than a very few of the most common species; and that when they hear a note that pleases them they set it down to one of the half-dozen or three or four songsters whose names they remember. I met with an amusing instance of this common mistake at a spot in the west of England, where I visited a castle on a hill, and was shown over the beautiful but steep grounds by a stout old dame, whose breath and temper were alike short. It was a bright morning in May, and the birds were in full song. As we walked through the shrubbery a blackcap burst into a torrent of wild heart-enlivening melody from amidst the foliage not more than three yards away. "How well that blackcap sings!" I remarked. "That blackbird," she corrected; "yes, it sings well."

She stuck to it that it was a blackbird, and to prove that I was wrong assured me that there were no blackcaps there. Finding that I refused to acknowledge myself in error, she got cross and dropped into sullen silence; but ten or fifteen minutes later she returned of her own accord to the subject.

"I've been thinking, sir," she said, "that you must be right. I said there are no blackcaps here because I've been told so, but all the same I've often remarked that the blackbird has two different songs. Now I know, but I'm so sorry that I didn't know a few days sooner." I asked her why. She replied, "The other day a young American lady came to the castle and I took her over the grounds. The birds were singing the same as to-day, and the young lady said, 'Now, I want you to tell me which is the blackcap's song.

Just think,' she said, 'what a distance I have come, from America! Well, when I was bidding good-bye to my friends at home I said, "Don't you envy me? I'm going to Old England to hear the blackcap's song."' Well, when I told her we had no blackcaps she was so disappointed; and yet, sir, if what you say is right, the bird was singing near us all the time!"

Poor young lady from America! I should have liked to know whose written words first fired her brain with desire of the blackcap's song--a golden voice in imagination's ear, while the finest home voices were merely silvern. I think of my own case; how in boyhood this same bird first warbled to me in some lines of a poem I read; and how, long years afterwards, I first heard the real song--beautiful, but how unlike the song I had imagined!

--one bright evening in early May, at Netley Abbey.

But the poet's name had meanwhile slipped out of memory; nothing but a vague impression remained (and still persists) that he flourished and had great fame about the beginning of the nineteenth century, and that now his (or her) fame and works are covered with oblivion.

To return to the subject of this paper: the wood wren--the secret of its charm. We see that, tried by ordinary standards, many other singers are its superiors; what, then, is the mysterious something in its music that makes it to some of us even better than the best? Speaking for myself, I should say because it is more harmonious, or in more perfect accord with the nature amid which it is heard; it is the truer woodland voice.

The chaffinch as a rule sings in open woods and orchards and groves when there is light and life and movement; but sometimes in the heart of a deep wood the silence is broken by its sudden loud lyric: it is unexpected and sounds unfamiliar in such a scene; the wonderfully joyous ringing notes are like a sudden flood of sunshine in a shady place. The sound is intensely distinct and individual, in sharp contrast to the low forest tones: its effect on the ear is similar to that produced on the sight by a vivid contrast in colours, as by a splendid scarlet or shining yellow flower blooming solitary where all else is green. The effect produced by the wood wren is totally different; the strain does not contrast with, but is complementary to, the "tremulous cadence low" of inanimate nature in the high woods, of wind-swayed branches and pattering of rain and lisping and murmuring of innumerable leaves--the elemental sounds out of which it has been fashioned. In a sense it may be called a trivial and a monotonous song--the strain that is like a long tremulous cry, repeated again and again without variation; but it is really beyond criticism--one would have to begin by depreciating the music of the wind. It is a voice of the beechen woods in summer, of the far-up cloud of green, translucent leaves, with open spaces full of green shifting sunlight and shadow.

Though resonant and far-reaching it does not strike you as loud, but rather as the diffused sound of the wind in the foliage concentrated and made clear--a voice that has light and shade, rising and passing like the wind, changing as it flows, and quivering like a wind-fluttered leaf. It is on account of this harmony that it is not trivial, and that the ear never grows tired of listening to it: sooner would it tire of the nightingale--its purest, most brilliant tone and most perfect artistry.

The continuous singing of a skylark at a vast height above the green, billowy sun and shadow-swept earth is an etherealised sound which fills the blue space, fills it and falls, and is part of that visible nature above us, as if the blue sky, the floating clouds, the wind and sunshine, has something for the hearing as well as for the sight. And as the lark in its soaring song is of the sky, so the wood wren is of the wood.

CHAPTER VI

THE SECRET OF THE WILLOW WREN

The willow wren is one of the commonest and undoubtedly the most generally diffused of the British songsters. A summer visitor, one of the earliest to arrive, usually appearing on the South Coast in the last week in March; a little later he may be met with in very nearly every wood, thicket, hedge, common, marsh, orchard, and large garden throughout the kingdom--it is hard to say, writes Seebohm, where he is not found. Wherever there are green perching-places, and small caterpillars, flies and aphides to feed upon, there you will see and hear the willow wren. He is a sweet and constant singer from the date of his arrival until about the middle of June, when he becomes silent for a season, resuming his song in July, and continuing it throughout August and even into September.

This late summer singing is, however, fitful and weak and less joyous in character than in the spring.

But in spite of his abundance and universality, and the charm of his little melody, he is not familiarly known to the people generally, as they know the robin redbreast, pied wagtail, dunnock, redstart, wheatear, and stonechat. The name we call him by is a very old one; it was first used in English by Ray, in his translation of Willughby's *Ornithology*, about three centuries ago; but it still remains a book-name unknown to the rustic. Nor has this common little bird any widely known vernacular name. If by chance you find a country-man who knows the bird, and has a name for it, this will be one which is applied indiscriminately to two, three, or four species. The willow wren, in fact, is one of those little birds that are "seen rather than distinguished," on account of its small size, modest colouring, and its close resemblance to other species of warblers; also on account of the quiet, gentle character of its song, which is little noticed in the spring and summer concert of loud, familiar voices.

One day in London during the late summer I was amused and at the same time a little disgusted at this general indifference to the delicate beauty in a bird-sound which distinguishes the willow wren even among such delicate singers as the warblers: it struck me as a kind of aesthetic hardness of hearing. I heard the song in the flower walk, in Kensington Gardens, on a Sunday morning, and sat down to listen to it; and for half an hour the bird continued to repeat his song two or three times a minute on the trees and bushes within half a dozen yards of my seat. Just after I had sat down, a throstle, perched on the topmost bough of a thorn that projected over the walk, began his song, and continued it a long time, heedless of the people passing below. Now, I noticed that in almost every case the person approaching lifted his eyes to the bird above, apparently admiring the music, sometimes even pausing for a moment in his walk; and that when two or three came together they not only looked up, but made some remark about the beauty of the song. But from first to last not one of all the passers-by cast a look towards the tree where the willow wren was singing; nor was there anything to show that the sound had any attraction for them, although they must have heard it. The loudness of the thrush prevented them from giving it any attention, and made it practically inaudible. It was like a pimpernel blossoming by the side of a poppy, or dahlia, or peony, where, even if seen, it would not be noticed as a beautiful flower.

In the chapter on the wood wren, I endeavoured to trace to its source the pleasurable feelings which the song of that bird produces in me and in many others--a charm exceeding that of many more celebrated vocalists. In that chapter the song of the willow wren was mentioned incidentally.

Now, these two--wood wren and willow wren--albeit nearly related, are, in the character of their notes, as widely different as it is possible for two songsters to be; and when we listen attentively to both, we recognise that the feeling produced in us differs in each case--that it has a different cause. In the case of the willow wren it might be said off-hand that our pleasure is simply due to the fact that it is a melodious sound, associated in our minds with summer scenes. As much could be said of any other migrant's song--nightingale, tree-pipit, blackcap, garden warbler, swallow, and a dozen more. But it does not explain the individual and very special charm of this particular bird--what I have ventured to call the secret of the willow wren. After all, it is not a deeply hidden secret, and has indeed been half guessed or hinted by various writers on bird melody; and as it also happens to be the secret of other singers besides the willow wren, we may, I think, find in it an explanation of the fact that the best singers do not invariably please us so well as some that are considered inferior.

The song of the willow wren has been called singular and unique among our birds; and Mr Warde Fowler, who has best described it, says that it forms an almost perfect cadence, and adds, "by which I mean that it descends gradually, not, of course, on the notes of our musical scale, by which no birds in their natural state would deign to be fettered, but through fractions of one or perhaps two of our tones, and without returning upward at the end." Now, this arrangement of its notes, although very rare and beautiful, does not give the little song its highest aesthetic value.

The secret of the charm, I imagine, is traceable to the fact that there is distinctly something human-like in the quality of the voice, its *timbre*. Many years ago an observer of wild birds and listener to their songs came to this country, and walking one day in a London suburb he heard a small bird singing among the trees. The trees were in an enclosure and he could not see the bird, but there would, he thought, be no difficulty in ascertaining the species, since it would only be necessary to describe its peculiar little song to his friends and they would tell him. Accordingly, on his return to the house he proceeded to describe the song and ask the name of the singer. No one could tell him, and much to his surprise, his account of the melody was received with smiles of amusement and incredulity. He described it as a song that was like a wonderfully bright and delicate human voice talking or laughingly saying something rather than singing. It was not until some time afterwards that the bird-lover in a strange land discovered that his little talker and laugher among the leaves was the willow wren. In vain he had turned to the ornithological works; the song he had heard, or at all events the song as he had heard it, was not described therein; and yet to this day he cannot hear it differently--cannot dissociate the sound from the idea of a fairy-like child with an exquisitely pure, bright, spiritual voice laughingly speaking in some green place.

And yet Gilbert White over a century ago had noted the human quality in the willow wren's voice when he described it as an "easy, joyous, laughing note." It is still better to be able to quote Mr Warde Fowler, when writing in *A Year with the Birds*, on the futile attempts which are often made to represent birds' songs by means of our notation, since birds are guided in their songs by no regular succession of intervals. Speaking of the willow wren in this connection, he adds: "Strange as it may seem, the songs of birds may perhaps be more justly compared with the human voice when speaking, than with a musical instrument, or with the human voice when singing."

The truth of this observation must strike any person who will pay close attention to the singing of birds; but there are two criticisms to be made on it. One is that the resemblance of a bird's song to a human voice when speaking is confined to some or to a few species; the second is that it is a mistake to think, as Mr Fowler appears to do, that the resemblance is wholly or mainly due to the fact that the bird's voice is free when singing--that, like the human voice in talking, it is not tied to tones and semitones. For instance, we note this peculiarity in the willow wren, but not in, say, the wren and chaffinch, although the songs of these two are just as free, just as independent of regular intervals as our voices when speaking and laughing. The resemblance in a bird's song to human speech is entirely due to the human-like quality in the voice; for we find that other songsters--notably the swallow--have a charm similar to that of the willow wren, although the notes of the former bird are differently arranged, and do not form anything like a cadence. Again, take the case of the blackbird. We are accustomed to describe the blackbird's voice as flute-like, and the flute is one of the instruments which most nearly resemble the human voice. Now, on account of the leisurely manner in which the blackbird gives out his notes, the resemblance to human speech is not so pronounced as in the case of the willow wren or swallow; but when two or three or half a dozen blackbirds are heard singing close together, as we sometimes hear them in woods and orchards where they are abundant, the effect is singularly beautiful, and gives the idea of a conversation being carried on by a set of human beings of arboreal habits (not monkeys) with glorified voices. Listening to these blackbird concerts, I have sometimes wondered whether or not they produced the same effect on others' ears as on mine, as of people talking to one another in high-pitched and beautiful tones. Oddly enough, it was only while writing this chapter that I by chance found an affirmative answer to my question. Glancing through Leslie's *Riverside Letters*, which I had not previously seen, I came upon the following remarks, quoted from Sir George Grove, in a letter to the author, on the blackbird's singing: "He selects a spot where he is within hearing of a comrade, and then he begins quite at leisure (not all in a hurry like the thrush) a regular conversation.

'And how are you? Isn't this a fine day? Let us have a nice talk,' etc., *etc.* He is answered in the same strain, and then replies, and so on. Nothing more thoughtful, more refined, more feeling, can be conceived." In another passage he writes: "I love them (the robins), but they fill a much smaller part than the blackbird does in my heart.

To hear the blackbird talking to his mate a field off, with deliberate, refined conversation, the very acme of grace and courtesy, is perfectly splendid."

There are two more common British songsters that produce much the same effect as the willow wren and blackbird; these are the swallow and pied wagtail. They are not in the first rank as melodists, and I can find no explanation of the fact that they please me better than the great singers other than their more human-like tones, which to my hearing have something of an exceedingly beautiful contralto sound. The swallow's song is familiar to every one, but that of the wagtail is not well known. The bird has two distinct songs: one, heard oftenest in early spring, consists of a low rambling warble, with some resemblance to the whinchat's song; it is the second song, heard occasionally until late June, frequently uttered on the wing--a torrent of loud, rapidly uttered, and somewhat swallow-like notes--that comes nearest in tone to the human voice, and has the greatest charm.

After these, we find other songsters with one or two notes, or a phrase, human-like in quality, in their songs. Of these I will only mention the blackcap, linnet, and tree-pipit. The most beautiful of the blackcap's notes, which come nearest to the blackbird, have this human sound; and certainly the most beautiful part of the linnet's song is the opening phrase, composed of notes that are both swallow-like and human-like.

It may appear strange to some readers that I put the tree-pipit, with his thin, shrill, canary-like pipe, in this list; but his notes are not all of this character; he is moreover a most variable singer; and it happens that in some individuals the concluding notes of the song have more of that peculiar human quality than any other British songster. No doubt it was a bird in which these human-like, languishing notes at the close of the song were very full and beautiful that inspired Burns to write his "Address to a Wood-lark."

The tree pipit is often called by that name in Scotland, where the true wood-lark is not found.

O stay, sweet warbling wood-lark, stay,
Nor quit for me the trembling spray,
A hopeless lover courts thy lay,
Thy soothing, fond complaining.

Again, again that tender part,
That I may catch thy melting art;
For surely that would touch her heart
Who kills me wi' disdaining.

Say, was thy little mate unkind,

And heard thee as the passing wind?
O nocht but love and sorrow joined
Sic notes o' wae could waken!

Thou tells o' never-ceasing care,
O' speechless grief and dark despair;
For pity's sake, sweet bird, nae mair,
Or my poor heart is broken!

Much more could be said about these and other species in the passerine order that have some resemblance, distinct or faint, to the human voice

in their singing notes--an echo, as it were, of our own common emotions, in most cases simply glad or joyous, but sometimes, as in the case of the tree-pipit, of another character. And even those species that are furthest removed from us in the character of the sounds they emit have some notes that suggest a highly brightened human voice. Witness the throstle and nightingale. The last approaches to the human voice in that rich, musical throb, repeated many times with passion, which is the invariable prelude to his song; and again, in that "one low piping note, more sweet than all," four times repeated in a wonderfully beautiful crescendo. Who that ever listened to Carlotta Patti does not remember sounds like these from her lips? It was commonly said of her that her voice was bird-like; certainly it was clarified and brightened beyond other voices--in some of her notes almost beyond recognition as a human voice.

It was a voice that had a great deal of the quality of gladness in it, but less depth of human passion than other great singers. Still, it was a human voice; and, just as Carlotta Patti (outshining the best of her sister-singers even as the diamond outsparkles all other gems) rose to the birds in her miraculous flights, so do some of the birds come down to and resemble us in their songs.

If I am right in thinking that it is the human note in the voices of some passerine birds that gives a peculiar and very great charm to their songs, so that an inferior singer shall please us more than one that ranks high, according to the accepted standard, it remains to ask why it should be so. Why, I mean, should the mere likeness to a human tone in a little singing-bird impart so great a pleasure to the mind, when the undoubtedly human-like voices of many non-passerine species do not as a rule affect us in the same way? As a matter of fact, we find in the multitude of species that resemble us in their voices a few, outside of the order of singers, that do give us a pleasure similar to that imparted by the willow wren, swallow, and tree-pipit. Thus, among British birds we have the wood-pigeon, and the stock-dove; the green woodpecker, with his laugh-like cry; the cuckoo, a universal favourite on account of his double fluty call; and (to those who are not inclined to be superstitious) the wood-owl, a most musical night-singer; and the curlew, with, in a less degree, various other shore birds. But in a majority of the larger birds of all orders the effect produced is different, and often the reverse of pleasant. Or if such sounds delight us, the feeling differs in character from that produced by the melodious singer, and is mainly due to that wildness with which we are in sympathy expressed by such sounds. Human-like voices are found among the auks, loons, and grebes; eagles and falcons; cuckoos, pigeons, goatsuckers, owls, crows, rails, ducks, waders, and gallinaceous birds. The cries and shrieks of some among these, particularly when heard in the dark hours, in deep woods and marshes and other solitary places, profoundly impress and even startle the mind, and have given rise all the world over to numberless superstitious beliefs. Such sounds are supposed to proceed from devils, or from demons inhabiting woods and waters and all desert places; from night-wandering witches; spirits sent to prophesy death or disaster; ghosts of dead men and women wandering by night about the world in search of a way out of it; and sometimes human beings who, burdened with dreadful crimes or irremediable griefs, have been changed into birds. The three British species best known on account of their supernatural character have very remarkable voices with a human sound in them: the raven with his angry, barking cry, and deep, solemn croak; the booming bittern; and the white or church owl, with his funereal screech.

It is, I think, plain that the various sensations excited in us by the cries, moans, screams, and the more or less musical notes of different species, are due to the human emotions which they express or seem to express. If the voice simulates that of a maniac, or of a being tortured in body or mind, or overcome with grief, or maddened with terror, the blood-curdling and other sensations proper to the occasion will be experienced; only, if we are familiar with the sound or know its cause, the sensation will be weak. Similarly, if in some deep, silent wood we are suddenly startled by a loud human whistle or shouted "Hi!" although we may know that a bird, somewhere in that waste of foliage around us, uttered the shout, we yet cannot help experiencing the feelings--a combination of curiosity, amusement, and irritation--which we should have if some friend or some human being had hailed us while purposely keeping out of sight.

Finally, if the bird-sounds resemble refined, bright, and highly musical human voices, the voices, let us say, of young girls in conversation, expressive of various beautiful qualities--sympathy, tenderness, innocent mirth, and overflowing gladness of heart--the effect will be in the highest degree delightful.

Herbert Spencer, in his account of the origin of our love of music in his *Psychology*, writes: "While the tones of anger and authority are harsh and coarse, the tones of sympathy and refinement are relatively gentle and of agreeable timbre.

That is to say, the timbre is associated in experience with the receipt of gratification, has acquired a pleasure-giving quality, and consequently the tones which in music have an allied timbre become pleasure-giving and are called beautiful. Not that this is the sole cause of their pleasure-giving quality.... Still, in recalling the tones of instruments which approach the tones of the human

voice, and observing that they seem beautiful in proportion to their approach, we see that this secondary aesthetic element is important."

As with instruments, so it is with bird voices; in proportion as they approach the tones of the human voice, expressive of sympathy, refinement, and other beautiful qualities, they will seem beautiful--in some cases even more beautiful than those which, however high they may rank in other ways, are yet without this secondary aesthetic element.

CHAPTER VII

SECRET OF THE CHARM OF FLOWERS

When my mind was occupied with the subject of the last chapter--the human quality in some sweet bird voices--it struck me forcibly that all resemblances to man in the animal and vegetable worlds and in inanimate nature, enter largely into and strongly colour our aesthetic feelings. We have but to listen to the human tones in wind and water, and in animal voices; and to recognise the human shape in plant, and rock, and cloud, and in the round heads of certain mammals, like the seal; and the human expression in the eyes, and faces generally, of many mammals, birds and reptiles, to know that these casual resemblances are a great deal to us. They constitute the *expression* of numberless natural sights and sounds

with which we are familiar, although in a majority of cases the resemblance being but slight, and to some one quality only, we are not conscious of the cause of the expression.

It was principally with flowers, which excite more attention and give more pleasure than most natural objects, that my mind was occupied in this connection; for here it seemed to me that the effect was similar to that produced on the mind by sweet human-like tones in bird music. In other words, a very great if not the principal charm of the flower was to be traced to the human associations of its colouring; and this was, in some cases, more than all its other attractions, including beauty of form, purity and brilliance of colour, and the harmonious arrangement of colours; and, finally, fragrance, where such a quality existed.

We see, then, that there is an intimate connection between the two subjects--human associations in the colouring of flowers and in the voices of birds; and that in both cases this association constitutes, or is a principal element in, the *expression*.

This connection, and the fact that the present subject was suggested and appeared almost an inevitable outcome of the one last discussed, must be my excuse for introducing a chapter on flowers in a book on birds--or birds and man. But an excuse is hardly needed. It must strike most readers that a great fault of books on birds is, that there is too much about birds in them, consequently that a chapter about something else, which has not exactly been dragged in, may come as a positive relief.

As the word expression which occurs with frequency in this chapter was not understood in the sense in which I used it on the first appearance of the book, it may be well to explain that it is not used here in its ordinary meaning as the quality in a face, or picture, or any work of art, which indicates thought or feeling. Here the word has the meaning given to it by writers on the aesthetic sense as descriptive of the quality imparted to an object by its associations. These may be untraceable: we may not be conscious and as a rule we are not conscious that any such associations exist; nevertheless they are in us all the time, and with what they add to an object may enhance and even double its intrinsic beauty and charm.

*

I have somewhere read a very ancient legend, which tells that man was originally made of many materials, and that at the last a bunch of wild flowers was gathered and thrown into the mixture to give colour to his eyes. It is a pretty story, but might have been better told, since it is certain that flowers which have delicate and beautiful flesh-tints are attractive mainly on that account, just as blue and some purples delight us chiefly because of their associations with the human iris.

The skin, too, needed some beautiful colour, and there were red as well as blue flowers in the bunch; and the red flowers being most abundant in nature and in greater variety of tints, give us altogether more pleasure than their beautiful rivals in our affection.

The blue flower is associated, consciously or not, with the human blue eye; and as the floral blue is in all or nearly all instances pure and beautiful, it is like the most beautiful human eye. This association, and not the colour itself, strikes me as the true cause of the superior attraction which the blue flower has for most of us. Apart from association blue is less attractive than red, orange, and yellow, because less luminous; furthermore green is the least effective background for such a colour as blue in so small an object as a flower; and, as a fact, we see that at a little distance the blue of the flower is absorbed and disappears in the surrounding green, while reds and yellows keep their splendour. Nevertheless the blue has a stronger hold on our affections. As a human colour, blue comes first in a blue-eyed race because it is the colour of the most important feature, and, we may say, of the very soul in man.

Some purple flowers stand next in our regard on account of their nearness in colour to the pure blue. The wild hyacinth, blue-bottle, violet, and pansy, and some others, will occur to every one.

These are the purple flowers in which blue predominates, and on that account have the same

expression as the blue. The purples in which red predominates are akin in expression to the reds, and are associated with flesh-tints and blood.

And here it may be noted that the blue and blue-purple flowers, which have the greatest charm for us, are those in which not only the colour of the eye but some resemblance in their form to the iris, with its central spot representing the pupil, appears.

For example, the flax, borage, blue geranium, periwinkle, forget-me-not, speedwell, pansy and blue pimpernel, are actually more to us than some larger and handsomer blue flowers, such as the blue-bottle, vipers' bugloss, and succory, and of blue flowers seen in masses.

With regard to the numerous blue and purple-blue flowers which we all admire, or rather for which we all feel so great an affection, we find that in many cases their very names have been suggested by their human associations--by their *expression*.

Love-in-a-mist, angels' eyes, forget-me-not, and heartsease, are familiar examples. Heartsease and pansy both strike us as peculiarly appropriate to one of our commonest and most universal garden flowers; yet we see something besides the sympathetic and restful expression which suggested these names in this flower--a certain suggestion of demureness, in fact, reminding those who have seen Guido's picture of the "Adoration of the Virgin," of one of his loveliest angels whose angelical eyes and face reveal some desire for admiration and love in the spectator. And that expression, too, of the pansy named Love-in-Idleness, has been described, coarsely or rudely it may be, in some of its country names: "Kiss me behind the garden gate," and, better (or worse) still, "Meet-her-i'-th'-entry-kiss-her-i'-th'-buttery." Of this order of names are None-so-pretty and Pretty maids, Pretty Betsy, Kiss-me-quick. Even such a name as Tears of the blood of Christ does not sound extravagantly fanciful or startling when we look at the glowing deep golden crimson of the wall flower; nor of a blue flower, the germander speedwell, such names as The more I see you the more I love you, and Angels' tears, and Tears of Christ, with many more.

A writer on our wild flowers, in speaking of their vernacular names of this kind, has said: "Could we penetrate to the original suggestive idea that called forth its name, it would bring valuable information about the first openings of the human mind towards nature; and the merest dream of such a discovery invests with a strange charm the words that could tell, if we could understand, so much of the forgotten infancy of the human race."

What a roll of words and what a mighty and mysterious business is here made of a very simple little matter! It is a charming example of the strange helplessness, not to say imbecility, which affects most of those who have been trained in our mind-killing schools; trained not to think, but taught to go for anything and everything they desire to know to the books. If the books in the British Museum fail to say why our ancestors hundreds of years ago named a flower None-so-pretty or Love-in-a-mist, why then we must be satisfied to sit in thick darkness with regard to this matter until some heaven-born genius descends to illuminate us! Yet I daresay there is not a country child who does not occasionally invent a name for some plant or creature which has attracted his attention; and in many cases the child's new name is suggested by some human association in the object--some resemblance to be seen in form or colour or sound. Not books but the light of nature, the experience of our own early years, the look which no person not blinded by reading can fail to see in a flower, is sufficient to reveal all this hidden wonderful knowledge about the first openings of the heart towards nature, during the remote infancy of the human race.

From this it will be seen that I am not claiming a discovery; that what I have called a secret of the charm of flowers is a secret known to every man, woman, and child, even to those of my own friends who stoutly deny that they have any such knowledge. But I think it is best known to children.

What I am here doing is merely to bring together and put in form certain more or less vague thoughts and feelings which I (and therefore all of us) have about flowers; and it is a small matter, but it happens to be one which no person has hitherto attempted.

It may be that in some of my readers' minds--those who, like the sceptical friends I have mentioned, are not distinctly conscious of the cause or secret of the expression of a flower--some doubt may still remain after what has been said of the blue and purple-blue blossom. Such a doubt ought to disappear when the reds are considered, and when it is found that the expression peculiar to red flowers varies infinitely in degree, and is always greatest in those shades of the colour which come nearest to the most beautiful flesh-tints.

When I say "beautiful flesh-tints" I am thinking of the aesthetic pleasure which we receive from the expression, the associations, of the red flower.

The expression which delights is in the soft and delicate shades; and in the texture which is sometimes like the beautiful soft skin; but the *expression* would exist still in the case of floral tints resembling the unpleasant reds, or the reds which disgust us, in the human face. And we most of us know that these distressing hues are to be seen in some flowers. I remember that I once went into a florist's shop, and seeing a great mass of hard purple-red cinerarias on a shelf I made some remark about them. "Yes, are they not beautiful?" said the woman in the shop. "No, I loathe the sight of them," I returned. "So do I!" she said very quickly, and then added that she called them beautiful because she had to sell them. She, too, had no doubt seen that same purple-red colour in the evil flower called "grog-blossom," and in the faces of many middle-aged lovers of the bottle, male and female, who would perish before their time, to the great relief of their kindred, and whose actions after they were gone would not smell sweet and blossom in the dust.

The reds we like best in flowers are the delicate roseate and pinky shades; they are more to us than the purest and most luminous tints. And here, as with bird notes which delight us on account of their resemblance to fresh, young, highly musical human voices, flowers please us best when they exhibit the loveliest human tints--the apple blossom and the bindweed, musk mallow and almond and wild rose, for example. After these we are most taken with the deeper but soft and not too luminous reds--the red which we admire in the red horse-chestnut blossom, and many other flowers, down

to the minute pimpernel. Next come the intense rosy reds seen in the herb-robert and other wild geraniums, valerian, red campion and ragged robin; and this shade of red, intensified but still soft, is seen in the willow-herb and foxglove, and, still more intensified, in the bell-and small-leafed heath. Some if not all of these pleasing reds have purple in them, and there are very many distinctly purple flowers that appeal to us in the same way that red flowers do, receiving their expression from the same cause. There is some purple colour in most skins, and even some blue.

The azured harebell, like thy veins,

is a familiar verse from *Cymbeline*; any one can see the resemblance to the pale blue of that admired and loved blossom in the blue veins of any person with a delicate skin. Purples and purplish reds in masses are mostly seen in young persons of delicate skins and high colour in frosty weather in winter, when the eyes sparkle and the face glows with the happy sensations natural to the young and healthy during and after outdoor exercise. The skin purples and purple-reds here described are beautiful, and may be matched to a nicety in many flowers; the human purple may be seen (to name a very common wild flower) in purple loosestrife and the large marsh mallow, and in dozens and scores of other familiar purple flowers; and the purple-red hue in many richly coloured skins has its exact shade in common hounds' tongue, and in other dark and purple-red flowers. But we always find, I fancy, that the expression due to human association in a purple flower is greatest when this colour (as in the human face) is placed side by side or fades into some shade of red or pink. I think we may see this even in a small flower like the fumitory, in which one portion is deep purple and all the rest of the blossoms a delicate pink. Even when the red is very intense, as in the common field poppy, the pleasing expression of purple on red is very evident.

To return to pure reds. We may say that just as purples in flowers look best, or have a greater degree of expression, when appearing in or with reds, so do the most delicate rose and pink shades appeal most to us when they appear as a tinge or blush on white flowers. Probably the flower that gives the most pleasure on account of its beautiful flesh-tints of different shades is the Gloire de Dijon rose, so common with us and so universal a favourite. Roses, being mostly of the garden, are out of my line, but they are certainly glorious to look at--glorious because of their associations, their expression, whether we know it or not. One can forgive Thomas Carew the conceit in his lines--

Ask me no more where Jove bestows
When June is past, the fading rose,
For in your beauty's orient deep
These flowers as in their causes sleep.

But all reds have something human, even the most luminous scarlets and crimsons--the scarlet verbena, the poppy, our garden geraniums, etc.--although in intensity they so greatly surpass the brightest colour of the lips and the most vivid blush on the cheek. Luminous reds are not, however, confined to lips and cheeks: even the fingers when held up before the eyes to the sun or to firelight show a very delicate and beautiful red; and this same brilliant floral hue is seen at times in the membrane of the ear. It is, in fact, the colour of blood, and that bright fluid, which is the life, and is often spilt, comes very much into the human associations of flowers. The Persian poet, whose name is best left unwritten, since from hearing it too often most persons are now sick and tired of it, has said,

I sometimes think that never blooms so red
The rose as where some buried Caesar bled.

There is many and many a "plant of the blood of men." Our most common Lovelies-bleeding with its "dropping wells" of crimson serves to remind us that there are numberless vulgar names that express this resemblance and association.

The thought or fancy is found everywhere in poetic literature, in the fables of antiquity, in the tales and folk-lore of all nations, civilised and barbarous.

I think that we can more quickly recognise this human interest in a flower, due to its colour, and best appreciate its aesthetic value from this cause, when we turn from the blues, purples, and reds, to the whites and the yellows. The feeling these last give us is distinctly different in character from that produced by the others. They are not like us, nor like any living sentient thing we are related to: there is no kinship, no human quality.

When I say "no kinship, no human quality,"

I refer to flowers that are entirely pure white or pure yellow; in some dull or impure yellows, and in white and yellow flowers that have some tinge or mixture of red or purple, we do get the expression of the red and purple flower. The crystalline and snow white of the whitest flowers do indeed resemble the white of the eyeballs and the teeth in human faces; but we may see that this human white colour by itself has no human association in a flower.

The whiteness of the white flower where there is any red is never unhuman, probably because a very brilliant red or rose colour on some delicate skins causes the light flesh-tints to appear white by contrast, and is the complexion known as "milk and roses." The apple-blossom is a beautiful example, and the beloved daisy--the "wee, modest, crimson-tipped flower," which would be so much less dear but for that touch of human crimson. This is the herb-Margaret of so many tender and pretty legends, that has white for purity and red for repentance. Even those who have never read these legends and that prettiest, most pathetic of all which tells of the daisy's origin, find a secret charm in the flower. Among other common examples are the rosy-white hawthorn, wood anemone, bindweed, dropwort, and many others. In the dropwort the rosy buds are seen among the creamy white open flowers; and the expression is always very marked and beautiful when there is any red or purple tinge or blush on cream-whites and ivory-whites. When we look from the dropwort to its nearest relative, the common meadow-sweet, we see how great a charm the touch of rose-red has given to the first: the meadow-sweet has no expression of the kind we are considering--no human association.

In pure yellow flowers, as in pure white, human interest is wanting. It is true that yellow is a human colour, since in the hair we find yellows of different shades--it is a pity that we cannot find, or have not found, a better word than "shades"

for the specific differences of a colour. There is the so-called tow, the tawny, the bronze, the simple yellow, and the golden, which includes many varieties, and the hair called carroty. But none of these has the flower yellow. Richard Jefferies tells us that when he placed a sovereign by the side of a dandelion he saw how unlike the two colours were--that, in fact, no two colours could seem more unlike than the yellow of gold and the yellow of the flower. It is not necessary to set a lock of hair and any yellow flower side by side to know how utterly different the hues are. The yellow of the hair is like that of metals, of clay, of stone, and of various earthy substances, and like the fur of some mammals, and like xanthophyll in leaf and stalk, and the yellow sometimes seen in clouds. When Ossian, in his famous address to the sun, speaks of his yellow hair floating on the eastern clouds, we instantly feel the truth as well as beauty of the simile. We admire the yellow flower for the purity and brilliance of its colour, just as we admire some bird notes solely for the purity and brightness of the sound, however unlike the human voice they may be. We also admire it in many instances for the exquisite beauty of its form, and the beauty of the contrast of pure yellow and deep green, as in the yellow flag, mimulus, and numerous other plants. But however much we may admire, we do not experience that intimate and tender feeling which the blues and reds inspire in us; in other words, the yellow flower has not the expression which distinguishes those of other colours. Thus, when Tennyson speaks of the "speedwell's darling blue," we know that he is right--that he expresses a feeling about this flower common to all of us; but no poet would make so great, so absurd a mistake as to describe the purest and loveliest yellow of the most prized and familiar wild flower--buttercup or kingcup, yellow flag, sea poppy, marsh marigold, or broom, or furze, or rock-rose, let us say--by such a word--the word that denotes an intimate and affectionate feeling--the feeling one cherishes for the loved ones of our kind. Nor could that word of Tennyson be properly used of any pure white flower--the stitchwort for instance; nor of any white and yellow flower like the Marguerite. But no sooner do you get a touch of rose or crimson in the whitest flower, as we see in the daisy and eyebright, than you can say of it that it is a "dear" or a "darling"

colour, and no one can find fault with the expression.

When we consider the dull and impure yellows sometimes seen in flowers, and some soft yellows seen in combination with pleasing wholesome reds, as in the honeysuckle, we may find something of the expression--the human association--in yellow flowers. For there *is* yellow in the skin, even in perfect health; it appears strongest on the neck, and spread round to the throat and chin, and is a warm buff, very beautiful in some women; but very little of this tint appears in the face.

When a tinge of this warm buffy yellow and creamy yellow is seen mixed with warmer reds, as in the Gloire de Dijon rose, the effect is most beautiful and the expression most marked. But the expression in flowers of a pale dull, impure yellow, where there is an expression, is unpleasant. It is the yellow of unhealthy skins, of faces discoloured by jaundice, dyspepsia, and other ailments. We commonly say of such flowers that they are "sickly"

in colour, and the association is with sick and decaying humanity. Gerarde, in describing such hues in flowers, was fond of the word "overworn"; and it was a very good word, and, like the one now in use, is derived from the association.

It will be noted by those who are acquainted with many flowers that I have given the names of but few--it may be too few--as examples, and that these are nearly all of familiar wild flowers.

My reason for not going to the garden is, that our cultivated blooms are not only artificially produced, and in some degree monstrosities, but they are seen in unnatural conditions, in crowds and masses, the various kinds too near together, and in most cases selected on account of their gorgeous colouring. The effect produced, however delightful it may be in some ways, is confusing to those simple natural feelings which flowers in a state of nature cause in us.

I confess that gardens in most cases affect me disagreeably; hence I avoid them, and think and know little about garden flowers. It is of course impossible not to go into gardens. The large garden is the greatly valued annexe of the large house, and is as much or more to the mistress than the coverts to the master; and when I am asked to go into the garden to see and admire all that is there, I cannot say, "Madam, I hate gardens."

On the contrary, I must weakly comply and pretend to be pleased. And when going the rounds of her paradise my eyes light by chance on a bed of tulips, or scarlet geraniums, or blue larkspurs, or detested calceolarias or cinerarias--a great patch of coloured flame springing out of a square or round bed of grassless, brown, desolate earth--the effect is more than disagreeable: the mass of colour glares at and takes possession of me, and spreads itself over and blots out a hundred delicate and prized images of things seen that existed in the mind.

But I am going too far, and perhaps making an enemy of a reader when I would much prefer to have him (or her) for a friend.

I have named few flowers, and those all the most familiar kinds, because it seemed to me that many examples would have had a confusing effect on readers who do not intimately know many species, or do not remember the exact colour in each case, and are therefore unable to reproduce in their minds the exact *expression*--the feeling which every flower conveys. On the other hand, the reader who knows and loves flowers, who has in his mind the distinct images of many scores, perhaps of two or three hundreds of species, can add to my example many more from his own memory.

There is one objection to the explanation given here of the cause of the charm of certain flowers, which will instantly occur to some readers, and may as well be answered in advance. This view, or theory, must be wrong, a reader will perhaps say, because my own preference is for a yellow flower (the primrose or daffodil, let us say), which to me has a beauty and charm exceeding all other flowers.

The obvious explanation of such a preference would be that the particular flower preferred is intimately associated with recollections of a happy childhood, or of early life. The associations will have made it a flower among flowers, charged with a subtle magic, so that the mere sight or smell of it calls up beautiful visions before the mind's eye.

Every person bred in a country place is affected in this way by certain natural objects and odours; and I recall the case of Cuvier, who was always affected to tears by the sight of some common yellow flower, the name of which I have forgotten.

The way to test the theory is to take, or think of, two or three or half-a-dozen flowers that have no personal associations with one's own early life--that are not, like the primrose and daffodil in the foregoing instance, sacred flowers, unlike all others; some with and some without human colouring, and consider the feeling produced in each case on the mind. If any one will look at, say, a Gloire de Dijon rose (in some persons its mental image will serve as well as the object itself) and then at a perfect white chrysanthemum, or lily, or other beautiful white flower; then at a perfect yellow chrysanthemum, or an allamanda, and at any exquisitely beautiful orchid, that has no human colour in it, which he may be acquainted with, he will probably say: I admire these chrysanthemums and other flowers more than the rose;

they are most perfect in their beauty--I cannot imagine anything more beautiful; but though the rose is less beautiful and splendid, the admiration I have for it appears to differ somewhat in character--to be mixed with some new element which makes this flower actually more to me than the others.

That something different, and something more, is the human association which this flower has for us in virtue of its colour; and the new element--the feeling it inspires, which has something of tenderness and affection in it--is one and the same with the feeling which we have for human beauty.

*

The foregoing has been given here with a few alterations, mainly verbal, as it appeared originally: something now remains to be added.

When writing about the wild flowers of West Cornwall in a work on *The Land's End* (1908), I returned to the subject of the charm of flowers due to their human colouring, and will repeat here much of what was there said.

Some of the readers of my flower chapter were not convinced that I had made out my case: it came as a surprise to them, and in some instances they cherished views of their own which they did not want to give up. Thus, two of my critics, writing independently, expressed their belief that flowers are precious to us and seem more beautiful than they are, because they are absolutely unrelated to our human life with its passions, sorrows, and tragedies--because, looking at flowers, we are taken into, or have glimpses of, another and brighter world such as a disembodied spirit might find itself in. It was nothing more than a pretty fancy; but I had other more thoughtful critics, and during my correspondence with them I became convinced of a serious omission in my account of the blue flower, when I said that its expression was due to association with the blue eye in man. The strongest of my friendly adversaries informed me that any man can revel at will among his own personal feelings and associations; that these were a "kind of bloom on the intrinsic beauty of things"--a happy phrase! He then asks: "What does blue suggest to a sailor? Sometimes the sea, sometimes the sky, sometimes the Blue Peter; but if you ask him what does blue paint suggest he would say *mourning*, that being the colour of a ship's mourning. Dr Sutton always called blue *no colour,* because it was the colour of death, the sign of the withdrawal of life."

This was interesting but fails as an argument since it was taken for granted in the chapter that blue in a flower or anything else, and in fact any colour, possesses individual associations for every one of us, according to what we are, to the temper of our minds, to the conditions in which we exist, our vocation, our early life, and so on. Blue may suggest sea and sky and the Blue Peter to a sailor, and yet the blue flower have an expression due to its human association in him as in another.

But my critic dropped by chance into something better, when he went on to ask, "Why shouldn't the heaven's blue make us love flowers? It does in my case I know, and I can *feel* the different blues of skies and air and distance in flower blue."

Undoubtedly he was right; the blue sky, fair weather, the open air, was a suggestion of the blue flower. It amazed me to think of the years I had spent under blue skies and of all I had felt about blue flowers, without stumbling upon this very simple fact. So simple, so near to the surface that you no sooner hear it than you imagine you have always known it! It was impossible to look at blue flowers and not be convinced of its truth, especially when the flowers were spread over considerable areas, as when I looked at wild hyacinths in the spring woods, or followed the interminable blue band of the vernal squill on the west Cornish coast, or saw large arid tracts of land in Suffolk blue with viper's bugloss.

Oddly enough just after the letter containing this criticism had reached me, another correspondent who was also among my opponents, sent me this fine passage from the old writer Sir John Ferne, on azure in blazoning: "Which blew colour representeth the Aire amongst the elements, that of all the rest is the greatest favourer of life, as the only nurse and maintainer of spirits in any living creature. The colour blew is commonly taken from the blue skye which appeareth so often as the tempests be overblowne, and notes prosperous successe and good fortune to the wearer in all his affayres."

In conclusion, after having adopted this new idea, my view is still that the human association is the principal factor in the expression of the blue flower, or at all events in a majority of flowers that bloom more or less sparingly and are usually seen as single blooms, not as mere splashes of colour.

Such are the pansy, violet, speedwell, hairbell, lungwort, blue geranium, *etc.* It may be that in all flowers of this kind too an element in the expression is due to the fair-weather associations with the colour; but these associations must be very much stronger in the case of a blue flower always seen in masses and sheets of colour as the wild hyacinth. Among dark-eyed races the fair-weather associations would alone give the blue flower its expression. I shouldn't wonder, if some explorer with a curious mind would try to find out what savages feel about flowers, that he would discover in them a special regard for the blue flower.

CHAPTER VIII

RAVENS IN SOMERSET

Mr Warde Fowler in his *Summer Studies of Birds and Books* has a pleasant chapter on wagtails, in which he remarks incidentally that he does not care for the big solemn birds that please, or are dear to, "Mr Hudson." Their bigness disturbs and their solemnity oppresses him. They do not twitter and warble, and flit hither and thither, flirting their feathers, and with their dainty gracefulness and airy, fairy ways wind themselves round his heart.

Wagtails are quite big enough for him; they are, in fact, as big as birds should be, and so long as these charming little creatures abound in these islands he (Mr Fowler) will be content. Indeed, he goes so far as to declare that on a desert island, without a human creature to share its solitude with him, he would be happy enough if only wagtails were there to keep him company. Mr Fowler is not joking; he tells us frankly what he thinks and feels, and when we come to consider the matter seriously, as he wishes us to do, we discover that there is nothing astonishing in his confession--that his mental attitude is capable of being explained.

It is only natural, in an England from which most of the larger birds have been banished, that he should have become absorbed in observing and in admiration of the small species that remain; for we observe and study the life that is nearest to us, and seeing it well we are impressed by its perfection--the perfect correspondence that exists between the creature and its surroundings--by its beauty, grace, and other attractive qualities, as we are not impressed by the life which is at a distance, and of which we only obtain rare and partial glimpses.

These thoughts passed through my mind one cold, windy day in spring, several hours of which I spent lying on the short grass on the summit of a cliff, watching at intervals a pair of ravens that had their nest on a ledge of rock some distance below. Big and solemn, and solemn and big, they certainly were, and although inferior in this respect to eagle, pelican, bustard, crane, vulture, heron, stork, and many another feathered notable, to see them was at the same time a pleasure and a relief. It also occurred to me at the time that, alone on a desert island, I should be better off with ravens than wagtails for companions; and this for an excellent reason. The wagtail is no doubt a very lively, pretty, engaging creature--so for that matter is the house fly--but between ourselves and the small birds there exists, psychologically, a vast gulf. Birds, says Matthew Arnold, live beside us, but unknown, and try how we will we can find no passages from our souls to theirs. But to Arnold--in the poem to which I have alluded at all events--a bird simply meant a caged canary; he was not thinking of the larger, more mammal-like, and therefore more human-like, mind of the raven, and, it may be added, of the crows generally.

The pair I spent so long a time in watching were greatly disturbed at my presence on the cliff. Their anxiety was not strange, seeing that their nest is annually plundered in the interest of the "cursed collector," as Sir Herbert Maxwell has taught us to name the worst enemy of the rarer British birds.

The "worst," I say; but there is another almost if not quite as bad, and who in the case of some species is really worse. At intervals of from fifteen to twenty minutes they would appear overhead uttering their angry, deep croak, and, with wings outspread, seemingly without an effort on their parts allow the wind to lift them higher and higher until they would look no bigger than daws; and, after dwelling for a couple of minutes on the air at that great height, they would descend to the earth again, to disappear behind a neighbouring cliff. And on each occasion they exhibited that wonderful aerial feat, characteristic of the raven, and rare among birds, of coming down in a series of long drops with closed wings. I am inclined to think that a strong wind is necessary for the performance of this feat, enabling the bird to fall obliquely, and to arrest the fall at any moment by merely throwing out the wings.

At any rate, it is a fact that I have never seen this method of descent used by the bird in calm weather.

It is totally different to the tumbling down, as if wounded, of ravens when two or more are seen toying with each other in the air--a performance which is also practised by rooks and other species of the crow family. The tumbling feat is indulged in only when the birds are playing, and, as it would appear, solely for the fun of the thing; the feat I am describing has a use, as it enables the bird to come down from a great height in the air in the shortest time and with the least expenditure of force possible. With the vertical fall of a bird like the gannet on its prey we are not concerned here, but with the descent to earth of a bird soaring at a considerable height. Now, many birds when rushing rapidly down appear to close their wings, but they are never wholly closed; in some cases they are carried as when folded, but are slightly raised from the body; in other cases the wing is tightly pressed against the side, but the primaries stand out obliquely, giving the descending bird the figure of a barbed arrow-head. This may be seen in daws, choughs, pipits, and many other species.

The raven suddenly closes his outspread wings, just as a man might drop his arms to his sides, and falls head downwards through the air like a stone bird cast down from its pedestal; but he falls obliquely, and, after falling for a space of twenty or thirty or more feet, he throws out his wings and floats for a few seconds on the air, then falls again, and then again, until the earth is reached.

Let the reader imagine a series of invisible wires stretched, wire above wire, at a distance of thirty or forty yards apart, to a height of six or seven hundred yards from the earth. Let him next imagine an acrobat, infinitely more daring, more agile, and graceful in action than any performer he has ever seen, standing on the highest wire of all, in his black silk tights, against the blue sky, his arms outstretched; then dropping his arms to his sides and diving through the air to the next wire, then to the next, and so on successively until he comes to the earth. The feat would be similar, only on a larger scale and less beautiful than that of the ravens as I witnessed it again and again from the cliff on that windy day.

While watching this magnificent display it troubled me to think that this pair of ravens would probably not long survive to be an ornament to the coast.

Their nest, it has been stated, is regularly robbed, but I had been informed that in the summer of 1894

a third bird appeared, and it was then conjectured that the pair had succeeded in rearing one of their young. About a month later a raven was picked up dead on the coast by a boatman,--killed, it was believed, by his fellow-ravens,--and since then two birds only have been seen. There are only two more pair of ravens on the Somersetshire coast, and, as one of these has made no attempt to breed of late, we may take it that the raven population of this county, where the species was formerly common, has now been reduced to two pairs.

Anxious to find out if there was any desire in the place to preserve the birds I had been observing, I made many inquiries in the neighbourhood, and was told that the landlord cared nothing about them, and that the tenant's only desire was to see the last of them. The tenant kept a large number of sheep, and always feared, one of his men told me, that the ravens would attack and kill his lambs. It was true that they had not done so as yet, but they might kill a lamb at any time; and, besides, there were the rabbits--the place swarmed with them--there was no doubt that a young rabbit was taken occasionally.

Why, then, I asked, if they were so destructive, did not his master go out and shoot them at once?

The man looked grave, and answered that his master would not do the killing himself, but would be very glad to see it done by some other person.

How curious it is to find that the old superstitions about the raven and the evil consequences of inflicting wilful injury on the bird still survive, in spite of the fact that the species has been persecuted almost to extirpation!

"Have you not read, sir," Don Quixote is made to say, "the annals and histories of England, wherein are renowned and famous exploits of King Arthur, of whom there goes a tradition, and a common one, all over that kingdom of Great Britain, that the king did not die, but that by magic art he was transformed into a raven, and that in process of time he shall reign again and recover his kingdom and sceptre, for which reason it cannot be proved that, from that day to this, any Englishman has killed a raven?"

Now, it is certain that many Englishmen kill ravens, also that if the country people in England ever had any knowledge of King Arthur they have long forgotten it. Nevertheless this particular superstition still exists. I have met with it in various places, and found an instance of it only the other day in the Midlands, where the raven no longer breeds.

Near Broadway, in Worcestershire, there is a farm called "Kite's Nest," where a pair of ravens bred annually up to about twenty-eight or thirty years ago, when the young were taken and the nest pulled down by three young men from the village: to this day it is related by some of the old people that the three young men all shortly came to bad ends. Near Broadway an old farmer told me that since the birds had been driven away from "Kite's Nest" he had not seen a raven in that part of the country until one made its appearance on his farm about four years ago. He was out one day with his gun, cautiously approaching a rabbit warren, when the bird suddenly got up from the mouth of a burrow, and coming straight to him, hovered for some seconds above his head, not more than thirty yards from him. "It looked as if he wanted to be shot at," said the old man, "but he's no bird to be shot at by I.

"Twould be bad for I to hurt a raven, and no mistake."

Continuing my inquiries about the Somerset ravens, I found a man who was anxious that they should be spared. His real reason was that their eggs for him were golden eggs, for he lived near the cliff, and had an eye always on them, and had been successful for many years in robbing their nest, until he had at length come to look on these birds almost as his own property. Being his he loved them, and was glad to talk about them to me by the hour. Among other things he related that the ravens had for very near neighbours on the rocks a pair of peregrine falcons, and for several years there had always been peace between them. At length one winter afternoon he heard loud, angry cries, and presently two birds appeared above the cliff--a raven and a falcon--engaged in desperate battle and mounting higher and higher as they fought. The raven, he said, did not croak, but constantly uttered his harsh, powerful, barking cry, while the falcon emitted shrill, piercing cries that must have been audible two miles away.

At intervals as they rose, wheeling round and round, they struck at each other, and becoming locked together fell like one bird for a considerable distance; then they would separate and mount again, shrieking and barking. At length they rose to so great a height that he feared to lose sight of them; but the struggle grew fiercer; they closed more often and fell longer distances, until they were near the earth once more, when they finally separated, flying away in opposite directions. He was afraid that the birds had fatally injured each other, but after two or three days he saw them again in their places.

It was not possible for him, he told me, to describe the feelings he had while watching the birds. It was the most wonderful thing he had ever witnessed, and while the fight lasted he looked round from time to time, straining his eyes and praying that some one would come to share the sight with him, and because no one appeared he was miserable.

I could well understand his feeling, and have not ceased to envy him his good fortune. Thinking, after leaving him, of the sublime conflict he had described, and of the raven's savage nature, Blake's question in his "Tiger, tiger, burning bright" came to my mind: Did He who made the lamb make thee?

We can but answer that it was no other; that when the Supreme Artist had fashioned it with bold, free lines out of the blue-black rock, he smote upon it with his mallet and bade it live and speak; and its voice when it spoke was in accord with its appearance and temper--the savage, human-like croak, and the loud, angry bark, as if a deep-chested man had barked like a blood-hound.

How strange it seems, when we come to think of it, that the owners of great estates and vast parks, who are lovers of wild nature and animal life, and should therefore have been most anxious to preserve this bird, have allowed it to be extirpated! "A raven tree," says the author of the *Birds of Wiltshire*, "is no mean ornament to a park, and speaks of a wide domain and large timber, and an ancient family; for the raven is an aristocratic bird and cannot brook a confined property and trees of a young growth. Would that its predilection were more humoured and a secure retreat allowed it by the larger proprietors in the land!"

The wide domains, the large timber, and the ancient families survive, but the raven has vanished. It occasionally takes a young rabbit. But the human ravens of Somerset--to wit, the men and boys who have as little right to the rabbits--do the same. I do not suppose that in this way fewer than ten thousand to twenty thousand rabbits are annually "picked up," or "poached"--if any one likes that word better--in the county. Probably a larger number. The existence of a pair of ravens on an estate of twenty or thirty thousand acres would not add much to the loss. No doubt the raven kills other creatures that are preserved for sport, but it does not appear that its extermination has improved things in Somerset. Thirty years ago, when black-game was more plentiful than it is now, the raven was to be met with throughout the county, and was abundant on Exmoor and the Quantocks. The old head keeper on the Forest of Exmoor told me that when he took the place, twenty-five years ago, ravens, carrion crows, buzzards, and hawks of various kinds were very abundant, and that the war he had waged against them for a quarter of a century had well-nigh extirpated all these species. He had kept a careful record of all birds killed, noting the species in every case, as he was paid for all, but the reward varied, the largest sum being given for the largest birds--ravens and buzzards. His book shows that in one year, a quarter of a century ago, he was paid for fifty-two ravens shot and trapped. After that the number annually diminished rapidly, and for several years past not one raven had been killed.

At present one may go from end to end of the county, which is a long one, and find no raven; but in very many places, from North Devon to the borders of Gloucestershire, one would find accounts of "last ravens." Even in the comparatively populous neighbourhood of Wells at least three pairs of ravens bred annually down to about twenty years ago--one pair in the tower on Glastonbury Tor, one on the Ebor rocks, and one at Wookey Hole, two miles from the town.

But Somerset is no richer in memories of "last ravens" than most English counties. A selection of the most interesting of such memories of ravens expelled from their ancestral breeding-places during the last half-century would fill a volume. In conclusion I will give one of the raven stories I picked up in Somerset. It was related to me by Dr Livett, who has been the parish doctor in Wells for over sixty years, and was able to boast, before retiring in 1898, that he was the oldest parish doctor in the kingdom.

About the year 1841 he was sent for to attend a cottage woman at Priddy--a desolate little village high up in the Mendips, four or five miles from Wells.

He had to remain some hours at the cottage, and about midnight he was with the other members of the family in the living-room, when a loud tapping was heard on the glazed window. As no one in the room moved, and the tapping continued at intervals, he asked why some one did not open the door. They replied that it was only the ravens, and went on to tell him that a pair of these birds roosted every night close by, and invariably when a light was seen burning at a late hour in any cottage they would come and tap at the window. The ravens had often been seen doing it, and their habit was so well known that no notice was taken of it.

CHAPTER IX

OWLS IN A VILLAGE

In November, when tramping in the Midlands, I paid a visit to a friend who had previously informed me, in describing the attractions of the small, remote, rustic village he lived in, that it was haunted by owls.

The night-roving bird that inhabits the country village and its immediate neighbourhood is, in most cases, the white or barn owl, the owl that prefers a loft in a barn or a church tower for home and breeding-place to the hollow, ivied tree. The loft is dry and roomy, the best shelter from the storm and the tempest, although not always from the tempest of man's insensate animosity. The larger wood owl is supposed to have a different disposition, to be a dweller in deep woods, in love with "seclusion, gloom, and retirement,"--a thorough hermit. It is not so everywhere, certainly not in my friend's Gloucestershire village, where the white owl is unknown, while the brown or wood owl is quite common. But it is not a thickly wooded district; the woods there are small and widely separated. There is, however, a deal of old hedgerow timber and many large trees scattered about the fields. These the owl inhabits and is abundant simply because the gamekeeper is not there with his everlasting gun; while the farmers look on the bird rather as a friend than an enemy.

To go a little further into the matter, there are no gamekeepers because the landowners cannot afford the expensive luxury of hand-reared pheasants. The country is, or was, a rich one; but the soil is clay so extraordinarily stiff that four or five horses are needed to draw a plough. It is, indeed, strange to see five huge horses, all in line, dragging a plough, and moving so slowly that, when looked at from a distance, they appear not to move at all. If here and there a little wheat is still grown, it is only because, as the farmers say, "We mun have straw." The land has mostly gone out of cultivation, many vacant farms could be had at about five shillings an acre, and the landlords would in many cases, when pay day came round, be glad to take half a crown and forgive the rest.

The fields that were once ploughed are used for grazing, but the sheep and cattle on them are very few; one can only suppose that the land is not suitable for grazing purposes, or else that the farmers are too poor to buy sufficient stock.

Viewed from some eminence, the wide, green country appears a veritable waste; the idle hedges enclosing vacant fields, the ancient scattered trees, the absence of life, the noonday quiet, where the silence is only broken at intervals by some distant bird voice, strangely impress the mind as by a vision of a time to come and of an England dispeopled. It is restful; there is a melancholy charm in it similar to that of a nature untouched by man, although not so strong. Here, everywhere are visible the marks of human toil and ownership--the wave-like, parallel ridges in the fields, now mantled with grass, and the hedges that cut up the surface of the earth into innumerable segments of various shapes and sizes. It is not wild, but there is something in it of the desolaton that accompanies wildness--a promise soon to be fulfilled, now that grass and herbage will have freedom to grow, and the hedges that have been trimmed for a thousand years will no longer be restrained from spreading.

In this district the farmhouses and cottages are not scattered over the country. The farm-buildings, as a rule, form part of the village; the villages are small and mostly hidden from sight among embowering trees or in a coombe. From the high ground in some places it is possible to gaze over many miles of surrounding country and not see a human habitation; hours may sometimes be passed in such a spot without a human figure appearing in the landscape.

The village I was staying at is called Willersey; the nearest to it, a little over a mile away, is Saintbury.

This last was just such a pretty peaceful spot as would tempt a world-weary man to exclaim on first catching sight of it, "Here I could wish to end my days." A little old-world village, set among trees in the sheltering hollow of a deep coombe, consisting of thatched stone cottages, grouped in a pretty disorder; a modest ale-house; a parsonage overgrown with ivy; and the old stone church, stained yellow and grey with lichen, its low square tower overtopped by the surrounding trees. It was a pleasure merely to sit idle, thinking of nothing, on the higher part of the green slope, with that small centre of rustic life at my feet. For many hours of each day it was strangely silent, the hours during which the men were away at a distance in the fields, the children shut up in school, and the women in their cottages. An occasional bird voice alone broke the silence--the distant harsh call of a crow, or the sudden startled note of a magpie close at hand, a sound that resembles the broken or tremulous bleat of a goat. If an apple dropped from a tree in the village, its thud would be audible from end to end of the little crooked street in every cottage it would be known that an apple had dropped. On some days the sound of the threshing-machine would be heard a mile or two away; in that still atmosphere it was like the prolonged hum of some large fly magnified a million times. A musical sound, buzzing or clear, at times tremulous, rising or falling at intervals, it would swell and fill the world, then grow faint and die away. This is one of the artificial sounds which, like distant chimes, harmonise with rural scenes.

Towards evening the children were all at play, their shrill cries and laughter sounding from all parts of the village. Then, when the sun had set and the landscape grew dim, they would begin to call to one another from all sides in imitation of the wood owl's hoot. During these autumn evenings the children at this spot appeared to drop naturally into the owl's note, just as in spring in all parts of England they take to mimicking the cuckoo's call. Children are like birds of a social and loquacious disposition in their fondness for a set call, a penetrative cry or note, by means of which they can converse at long distances.

But they have no settled call of their own, no cry as distinctive as that of one of the lower animals. They mimic some natural sound. In the case of the children of these Midland villages it is the wood owl's clear prolonged note; and in every place where some animal with a striking and imitable voice is found its call is used by them. Where no such sound is heard, as in large towns, they invent a call; that is, one invents it and the others immediately take it up. It is curious that the human species, in spite of its long wild life in the past, should have no distinctive call, or calls, universally understood.

Among savage tribes the men often mimic the cry of some wild animal as a call, just as our children do that of an owl by night, and of some diurnal species in the daytime. Other tribes have a call of their own, a shout or yell peculiar to the tribe; but it is not used instinctively--it is a mere symbol, and is artificial, like the long-drawn piercing *coo-ee* of the Australian colonists in the bush, and the abrupt *Hi!* with which we hail a cab, with other forms of halooing; or even the lupine gurgled yowl of the morning milkman.

After dark the silence at the village was very profound until about half-past nine to ten o'clock, when the real owls, so easily to be distinguished from their human mockers, would begin their hooting--a single, long, uninflected note, and after it a silent interval of eight or ten seconds; then the succeeding longer, much more beautiful note, quavering at first, but growing steady and clear, with some slight modulation in it. The symbols *hoo-hoo* and *to-whit to-who*, as Shakespeare wrote it, stand for the wood owl's note in books; but you cannot spell the sound of an oaten straw, nor of the owl's pipe. There is no *w* in it, and no *h* and no *t*. It suggests some wind instrument that resembles the human voice, but a very un-English one--perhaps the high-pitched somewhat nasal voice of an Arab intoning a prayer to Allah.

One cannot hit on the precise instrument, there are so many; perhaps it is obsolete, and the owl was taught his song by lovers in the long ago, who wooed at twilight in a forgotten tongue,

And gave the soft winds a voice,
With instruments of unremembered forms.

No, that cannot be; for the wood owl's music is doubtless older than any instrument made by hands to be blown by human lips. Listening by night to their concert, the many notes that come from far and near, human-like, yet airy, delicate, mysterious, one could imagine that the sounds had a meaning and a message to us; that, like the fairy-folk in Mr Yeats's Celtic lyric, the singers were singing--

We who are old, old and gay,
O, so old;
Thousands of years, thousands of years,
If all were told!

The fairies certainly have a more understandable way of putting it than the geologists and the anthropologists when we ask them to tell us how long it is since Palaeolithic man listened to the hooting of the wood owl. Has this sound the same meaning for us that it had for him--the human being that did not walk erect, and smile, and look on heaven, but went with a stoop, looking on the earth? No, and Yes.

Standing alone under the great trees in the dark still nights, the sound seems to increase the feeling of loneliness, to make the gloom deeper, the silence more profound. Turning our visions inward on such occasions, we are startled with a glimpse of the night-side of nature in the soul: we have with us strange unexpected guests, fantastic beings that are in no way related to our lives; dead and buried since childhood, they have miraculously been restored to life.

When we are back in the candlelight and firelight, and when the morrow dawns, these children of night and the unsubstantial appearance of things fade away

Into the light of common day.

The villagers of Saintbury are, however, still in a somewhat primitive mental condition; the light of common day does not deliver them from the presence of phantoms, as the following instance will show.

Near Willersey there is a group of very large old elm-trees which is a favourite meeting-place of the owls, and one very dark starless night, about ten o'clock, I had been listening to them, and after they ceased hooting I remained for half an hour standing motionless in the same place. At length, in the direction of Saintbury, I heard the dull sound of heavy stumbling footsteps coming towards me over the rough, ridgy field. Nearer and nearer the man came, until, arriving at the hedge close to which I stood, he scrambled through, muttering maledictions on the thorns that scratched and tore him; then, catching sight of me at a distance of two or three yards, he started back and stood still very much astonished at seeing a motionless human figure at that spot. I greeted him, and, to explain my presence, remarked that I had been listening to the owls.

"Owls!--listening to the owls!" he exclaimed, staring at me. After a while he added, "We have been having too much of the owls over at Saintbury."

Had I heard, he asked, about the young woman who had dropped down dead a week or two ago, after hearing an owl hooting near her cottage in the daytime?

Well, the owl had been hooting again in the same tree, and no one knew who it was for and what to expect next. The village was in an excited state about it, and all the children had gathered near the tree and thrown stones into it, but the owl had stubbornly refused to come out.

That about the young woman he had spoken of is a queer little story to read in this enlightened land.

She was apparently in very good health, a wife, and the mother of a small child; but a few weeks before her sudden death a strange thing occurred to trouble her mind. One afternoon, when sitting alone in her cottage taking tea, she saw a cricket come in at the open door, and run straight into the middle of the room. There it remained motionless, and without stirring from her seat she took a few moist tea-leaves and threw them down near the welcome guest. The cricket moved up to the leaves, and when it touched them and appeared just about to begin sucking their moisture, to her dismay it turned aside, ran away out at the door, and disappeared. She informed all her neighbours of this startling occurrence, and sadly spoke of an aunt who was living at another village and was known to be in bad health. "It must be for her," she said; "we'll soon be hearing bad news of her, I'm thinking." But no bad news came, and when she was beginning to believe that the strange cricket that had refused to remain in the house had proved a false prophet, the warning of the owl came to startle her afresh. At noonday she heard it hooting in the great horse-chestnut overgrown with ivy that stands at the roadside, close to her cottage.

The incident was discussed by the villagers with their usual solemnity and head-shakings, and now the young woman gave up all hopes of her sick aunt's recovery; for that one of her people was going to die was certain, and it could be no other than that ailing one. And, after all, the message and warning was for her and not the aunt. Not many days after the owl had hooted in broad daylight, she dropped down dead in her cottage while engaged in some domestic work.

On the following morning I went with the friend I was visiting at Willersey to Saintbury, and the story heard overnight was confirmed. The owl *had* been hooting in the daytime in the same old horse-chestnut tree from which it had a short time ago foretold the young woman's death. One of the villagers, who was engaged in repairing the thatch of a cottage close to the tree, informed us that the owl's hooting had not troubled him in the least.

Owls, he truly said, often hoot in the daytime during the autumn months, and he did not believe that it meant death for some one.

This sceptical fellow, it is hardly necessary to say, was a young man who had spent a good deal of his time away from the village.

At Willersey, a Mr Andrews, a lover of birds who owns a large garden and orchard in the village, gave me an entertaining account of a pet wood owl he once had. He had it as a young bird and never confined it. As a rule it spent most of the daylight hours in an apple loft, coming forth when the sun was low to fly about the grounds until it found him, when it would perch on his shoulder and spend the evening in his company. In one thing this owl differed from most pet birds which are allowed to have their liberty: he made no difference between the people of the house and those who were not of it; he would fly on to anybody's shoulder, although he only addressed his hunger-cry to those who were accustomed to feed him.

As he roamed at will all over the place he became well known to every one, and on account of his beauty and perfect confidence he grew to be something of a village pet. But short days with long, dark evenings--and how dark they can be in a small, tree-shaded, lampless village!--wrought a change in the public feeling about the owl. He was always abroad in the evening, gliding about unseen in the darkness on downy silent wings, and very suddenly dropping on to the shoulder of any person--man, woman, or child--who happened to be out of doors. Men would utter savage maledictions when they felt the demon claws suddenly clutch them; girls shrieked and fled to the nearest cottage, into which they would rush, palpitating with terror. Then there would be a laugh, for it was only the tame owl; but the same terror would be experienced on the next occasion, and young women and children were afraid to venture out after nightfall lest the ghostly creature with luminous eyes should pop down upon them.

At length, one morning the bird came not back from his night-wandering, and after two days and nights, during which he had not been seen, he was given up for lost. On the third day Mr Andrews was in his orchard, when, happening to pass near a clump of bushes, he heard the owl's note of recognition very faintly uttered. The poor bird had been in hiding at that spot the whole time, and when taken up was found to be in a very weak condition and to have one leg broken. No doubt one of the villagers on whose shoulders it had sought to alight, had struck it down with his stick and caused its injury. The bone was skilfully repaired and the bird tenderly cared for, and before long he was well again and strong as ever; but a change had come over his disposition. His confidence in his human fellow-creatures was gone; he now regarded them all--even those of the house--with suspicion, opening wide his eyes and drawing a little back when any person approached him. Never more did he alight on any person's shoulder, though his evenings were spent as before in flying about the village. Insensibly his range widened and he became wilder. Human companionship, no longer pleasant, ceased to be necessary; and at length he found a mate who was willing to overlook his pauper past, and with her he went away to live his wild life.

CHAPTER X

THE STRANGE AND BEAUTIFUL SHELDRAKE

At the head of the Cheddar valley, a couple of miles from the cathedral city of Wells, the Somerset Axe is born, gushing out noisily, a mighty volume of clear cold water, from a cavern in a black precipitous rock on the hillside. This cavern is called Wookey Hole, and above it the rough wall is draped with ivy and fern, and many small creeping plants and flowery shrubs rooted in the crevices; and in the holes in the rock the daws have their nests. They are a numerous and a vociferous colony, but the noise of their loudest cawings, when they rush out like a black cloud and are most excited, is almost drowned by the louder roar of the torrent beneath--the river's great cry of liberty and joy on issuing from the blackness in the hollow of the hills into the sunshine of heaven and the verdure of that beautiful valley. The Axe finishes its course fifteen miles away, for 'tis a short river, but they are pleasant miles in one of the fairest vales in the west of England, rich in cattle and in corn. And at the point where it flows into the Severn Sea stands Brean Down, a huge isolated hill, the last of the Mendip range on that side. It has a singular appearance: it might be likened in its form to a hippopotamus standing on the flat margin of an African lake, its breast and mouth touching the water, and all its body belly-deep in the mud; it is, in fact, a hill or a promontory united to the mainland by a strip of low flat land--a huge, oblong, saddle-backed hill projected into the sea towards Wales. Down at its foot, at the point where it touches the mainland, close to the mouth of the Axe, there is a farmhouse, and the farmer is the tenant of the entire hill, and uses it as a sheep-walk.

The sheep and rabbits and birds are the only inhabitants.

I remember a delightful experience I

had one cold windy but very bright spring morning near the farmhouse. There is there, at a spot where one is able to ascend the steep hill, a long strip of rock that looks like the wall of a gigantic ruined castle, rough and black, draped with ancient ivy and crowned with furze and bramble and thorn.

Here, coming out of the cold wind to the shelter of this giant ivy-draped black wall, I stood still to enjoy the sensations of warmth and a motionless air, when high above appeared a swift-moving little cloud of linnets, seemingly blown across the sky by the gale; but quite suddenly, when directly over me, the birds all came straight down, to drop like a shower of small stones into the great masses of ivy and furze and bramble. And no sooner had they settled, vanishing into that warm and windless greenery, than they simultaneously burst into such a concert of sweetest wild linnet music, that I was enchanted, and thought that never in all the years I had spent in the haunts of wild birds had I heard anything so fairy-like and beautiful.

On this hill, or down, at the highest point, you have the Severn Sea before you, and, beyond, the blue mountains of Glamorganshire, and, on the shore, the town of Cardiff made beautiful by distance, vaguely seen in the blue haze and shimmering sunlight like a dream city. On your right hand, on your own side of the narrow sea, you have a good view of the big young growing town of Weston-super-Mare--Bristol's Margate or Brighton, as it

has been called. It is built of Bath stone, and at this distance looks grey, darkened with the slate roofs, and a little strange; but the sight is not unpleasant, and if you wish to retain that pleasant impression, go not nearer to it than Brean Down, since on a closer view its aspect changes, and it is simply ugly. On your left hand you look over long miles, long leagues, of low flat country, extending to the Parret River, and beyond it to the blue Quantock range. That low land is on a level with the sea, and is the flattest bit of country in England, not even excepting the Ely district.

Apart from the charm which flatness has in itself for some persons--it has for me a very great charm on account of early associations--there is much here to attract the lover of nature. It is the chief haunt and paradise of the reed warbler, one of our sweetest songsters, and here his music may be heard amid more perfect surroundings than in any other haunt of the bird known to me in England.

This low level strip of country is mostly pasture-land, and is drained by endless ditches, full of reeds and sedges growing in the stagnant sherry-coloured water; dwarf hawthorn grows on the banks of the ditches, and is the only tree vegetation. Standing on one of the wide flat green fields or spaces, at a distance from the sandy dyke or ditch, it is strangely silent. Unless a lark is singing near, there is no sound at all; but it is wonderfully bright and fragrant where the green level earth is yellowed over with cowslips, and you get the deliciousness of that flower in fullest measure. On coming to the dyke you are no longer in a silent land with fragrance as its principal charm--you are in the midst of a perpetual flow and rush of sound. You may sit or lie there on the green bank by the hour and it will not cease; and so sweet and beautiful is it, that after a day spent in rambling in such a place with these delicate spring delights, on returning to the woods and fields and homesteads the songs of thrush and blackbird sound in the ear as loud and coarse as the cackling of fowls and geese.

It is in this district, from Brean Down westwards along the coast to Dunster, that I have been best able to observe and enjoy the beautiful sheldrake--almost the only large bird which is now permitted to exist in Somerset.

The sheldrake of the British Islands, called the common sheldrake (or sheld-duck) in the natural history books, for no good reason, since there is but one, is now becoming common enough as an ornamental waterfowl. It is to be seen in so many parks and private grounds all over the country that the sight of it in its conspicuous plumage must be pretty familiar to people generally. And many of those who know it best as a tame bird would, perhaps, say that the descriptive epithets of strange and beautiful do not exactly fit it. They would say that it has a striking appearance, or that it is peculiar and handsome in a curious way; or they might describe it as an abnormally slender and elegant-looking Aylesbury duck, whiter than that domestic bird, with a crimson beak and legs, dark-green glossy head, and sundry patches of chestnut-red and black on its snowy plumage. In calling it "strange" I was thinking of its manners and customs rather than of the singularity of its appearance.

As to its beauty, those who know it in a state of nature, in its haunts on the sea coast, will agree that it is one of the handsomest of our large wild birds. It cannot now be said that it is common, except in a few favoured localities. On the south coast it is all but extinct as a breeding species, and on the east side of England it is becoming increasingly rare, even in spots so well suited to it as Holy Island, and the coast at Bamborough Castle, with its great sand-hills. These same hills that look on the sea, and are greener than ivy with the everlasting green of the rough marram grass that

covers them, would be a very paradise to the sheldrake, but for man--vile man!--who watches him through a spy-glass in the breeding season to rob him of his eggs. The persecuted bird has grown exceedingly shy and cautious, but go he must to his burrow in the dunes, and the patient watcher sees him at a great distance on account of his conspicuous white plumage, and marks the spot, then takes his spade to dig down to the hidden eggs.

On the Somerset coast the bird is not so badly off, and I have had many happy days with him there. Simply to watch the birds at feed, when the tide goes out and they are busy searching for the small marine creatures they live on among the stranded seaweed, is a great pleasure. At such times they are most active and loquacious, uttering a variety of wild goose-like sounds, frequently rising to pursue one another in circles, or to fly up and down the coast in pairs, or strings of half a dozen birds, with a wonderfully graceful flight.

If, after watching this sea-fowl by the sea, a person will go to some park water to look on the same bird, pinioned and tame, sitting or standing, or swimming about in a quiet, listless way, he will be amazed at the difference in its appearance. The tame bird is no bigger than a domestic duck; the wild sheldrake, flying about in the strong sunshine, looks almost as large as a goose. A similar illusion is produced in the case of some other large birds.

Thus, the common buzzard, when rising in circles high above us, at times appears as big as an eagle, and it has been conjectured that this magnifying effect, which gives something of sublimity to the soaring buzzard, is caused by the sunlight passing through the semi-translucent wing and tail feathers.

In the case of the sheldrake, the exaggerated size may be an effect of strong sunlight on a flying white object. Seen on the wing at a distance the plumage appears entirely of a surpassing whiteness, the dark patches of chestnut, black, and deep green colour showing only when the bird is near, or when it alights and folds its white wings.

When the tide has covered their feeding-ground on the coast, the sheldrakes are accustomed to visit the low green pasture-lands, and may be seen in small flocks feeding like geese on the clover and grass. Here one day I saw about a dozen sheldrakes in the midst of an immense congregation of rooks, daws, and starlings feeding among some cows. It was a curious gathering, and the red Devons, shining white sheldrakes, and black rooks on the bright green grass, produced a singular effect.

Best of all it is to observe the birds when breeding in May. Brean Down is an ancient favourite breeding-site, and the birds breed there in the rabbit holes, and sometimes under a thick furze-bush on the ground. At another spot on this coast I have had the rare good fortune to find a number of pairs breeding at one spot on private enclosed land, where I could approach them very closely, and watch them any day for hours at a stretch, studying their curious sign-language, about which nothing, to my knowledge, has hitherto been written.

There were about thirty pairs, and their breeding-holes were mostly rabbit-burrows scattered about on a piece of sandy ground, about an acre and a half in extent, almost surrounded by water. When I watched them the birds were laying; and at about ten o'clock in the morning they would begin to come in from the sea in pairs, all to settle down at one spot; and by creeping some distance at the water-side among the rushes, I could get within forty yards of them, and watch them by the hour without being discovered by them. In an hour or so there would be forty or fifty birds forming a flock, each couple always keeping close together, some sitting on the short grass, others standing, all very quiet. At length one bird in the flock, a male, would all at once begin to move his head in a slow, measured manner from side to side, like a pianist swaying his body in time to his own music.

If no notice was taken of this motion by the duck sitting by his side dozing on the grass, the drake, would take a few steps forward and place himself directly before her, so as to compel her to give attention, and rock more vigorously than ever, haranguing her, as it were, although without words; the meaning of it all being that it was time for her to get up and go to her burrow to lay her egg. I do not know any other species in which the male takes it on himself to instruct his mate on a domestic matter which one would imagine to be exclusively within her own province; and some ornithologists may doubt that I have given a right explanation of these curious doings of the sheldrake. But mark what follows: The duck at length gets up, in a lazy, reluctant way, perhaps, and stretches a wing and a leg, and then after awhile sways *her* head two or three times, as if to say that she is ready. At once the drake, followed by her, walks off, and leads the way to the burrow, which may be a couple of hundred yards away; and during the walk she sometimes stops, whereupon he at once turns back and begins the swaying motion again. At last, arriving at the mouth of the burrow, he steps aside and invites her to enter, rocking himself again, and anon bending his head down and looking into the cavity, then drawing back again; and at last, after so much persuasion on his part, she lowers her head, creeps quietly down and disappears within. Left alone, the drake stations himself at the burrow's mouth, with head raised like a sentinel on duty; but after five or ten minutes he slowly walks back to the flock, and settles down for a quiet nap among his fellows.

They are all married couples; and every drake among them, when in some mysterious way he knows the time has come for the egg to be laid, has to go through the same long ceremonious performance, with variations according to his partner's individual disposition.

It is amusing to see at intervals a pair march off from the flock; and one wonders whether the others, whose turn will come by and by, pass any remarks; but the dumb conversation at the burrow's mouth is always most delightful to witness.

Sometimes the lady bird exhibits an extreme reluctance, and one can imagine her saying, "I have come thus far just to please you, but you'll never persuade me to go down into that horrid dark hole.

If I *must* lay an egg, I'll just drop it out here on the grass and let it take its chance."

It is rather hard on the drake; but he never loses his temper, never boxes her ears with his carmine red beak, or thrashes her with his shining white wings, nor does he tell her that she is just like a woman--an illogical fool. He is most gentle and considerate, full of distress and sympathy for her, and tells her again what he has said before, but in a different way; he agrees with her that it is dark and close down there away from the sweet sunlight, but that it is an old, old custom of the sheldrakes to breed in holes, and has its advantages; and that if she will only overcome her natural repugnance and fear of the dark, in that long narrow tunnel, when she is once settled down on the nest and feels the cold eggs growing warm again under her warm body she will find that it is not so bad after all.

And in the end he prevails; and bowing her pretty head she creeps quietly down and disappears, while he remains on guard at the door--for a little while.

CHAPTER XI

GEESE: AN APPRECIATION AND A MEMORY

One November evening, in the neighbourhood of Lyndhurst, I saw a flock of geese marching in a long procession, led, as their custom is, by a majestical gander; they were coming home from their feeding-ground in the forest, and when I spied them were approaching their owner's cottage.

Arrived at the wooden gate of the garden in front of the cottage, the leading bird drew up square before it, and with repeated loud screams demanded admittance. Pretty soon, in response to the summons, a man came out of the cottage, walked briskly down the garden path and opened the gate, but only wide enough to put his right leg through; then, placing his foot and knee against the leading bird, he thrust him roughly back; as he did so three young geese pressed forward and were allowed to pass in; then the gate was slammed in the face of the gander and the rest of his followers, and the man went back to the cottage. The gander's indignation was fine to see, though he had most probably experienced the same rude treatment on many previous occasions. Drawing up to the gate again he called more loudly than before; then deliberately lifted a leg, and placing his broad webbed foot like an open hand against the gate actually tried to push it open! His strength was not sufficient; but he continued to push and to call until the man returned to open the gate and let the birds go in.

It was an amusing scene, and the behaviour of the bird struck me as characteristic. It was this lofty spirit of the goose and strict adhesion to his rights, as well as his noble appearance and the stately formality and deliberation of his conduct, that caused me very long ago to respect and admire him above all our domestic birds. Doubtless from the aesthetic point of view other domesticated species are his superiors in some things: the mute swan, "floating double," graceful and majestical, with arched neck and ruffled scapulars; the oriental pea-fowl in his glittering mantle; the helmeted guinea-fowl, powdered with stars, and the red cock with his military bearing--a shining Elizabethan knight of the feathered world, singer, lover, and fighter. It is hardly to be doubted that, mentally, the goose is above all these; and to my mind his, too, is the nobler figure; but it is a very familiar figure, and we have not forgotten the reason of its presence among us. He satisfies a material want only too generously, and on this account is too much associated in the mind with mere flavours.

We keep a swan or a peacock for ornament; a goose for the table--he is the Michaelmas and Christmas bird. A somewhat similar debasement has fallen on the sheep in Australia. To the man in the bush he is nothing but a tallow-elaborating organism, whose destiny it is to be cast, at maturity, into the melting vat, and whose chief use it is to lubricate the machinery of civilisation. It a little shocks, and at the same time amuses, our Colonial to find that great artists in the parent country admire this most unpoetic beast, and waste their time and talents in painting it.

Some five or six years ago, in the *Alpine Journal*, Sir Martin Conway gave a lively and amusing account of his first meeting with A. D. M'Cormick, the artist who subsequently accompanied him to the Karakoram Himalayas. "A friend," he wrote, "came to me bringing in his pocket a crumpled-up water sketch or impression of a lot of geese. I was struck by the breadth of the treatment, and I remember saying that the man who could see such monumental magnificence in a flock of geese ought to be the kind of man to paint mountains, and render somewhat of their majesty."

I will venture to say that he looked at the sketch or impression with the artist's clear eye, but had not previously so looked at the living creature; or had not seen it clearly, owing to the mist of images--if that be a permissible word--that floated between it and his vision--remembered flavours and fragrances, of rich meats, and of sage and onions and sweet apple sauce. When this interposing mist is not present, who can fail to admire the goose--that stately bird-shaped monument of clouded grey or crystal white marble, to be seen standing conspicuous on any village green or common in England? For albeit a conquered bird, something of the ancient wild and independent spirit survives to give him a prouder bearing than we see in his fellow feathered servants. He is the least timid of our domestic birds, yet even at a distance he regards your approach in an attitude distinctly reminiscent of the grey-lag goose, the wariest of wild fowl, stretching up his neck and standing motionless and watchful, a sentinel on duty. Seeing him thus, if you deliberately go near him he does not slink or scuttle away, as other domestic birds of meaner spirits do, but boldly advances to meet and challenge you. How keen his senses are, how undimmed by ages of captivity the ancient instinct of watchfulness is in him, every one must know who has slept in lonely country houses. At some late hour of the night the sleeper was suddenly awakened by the loud screaming of the geese; they had discovered the approach of some secret prowler, a fox perhaps, or a thievish tramp or gipsy, before a dog barked. In many a lonely farmhouse throughout the land you will be told that the goose is the better watch-dog.

When we consider this bird purely from the aesthetic point of view--and here I am speaking of geese generally, all of the thirty species of the sub-family Anserinae, distributed over the cold and temperate regions of the globe--we find that several of them possess a rich and beautiful colouring, and, if not so proud, often a more graceful carriage than our domestic bird, or its original, the wild grey-lag goose. To know these birds is to greatly admire them, and we may now add that this admiration is no new thing on the earth. It is the belief of distinguished Egyptologists that a fragmentary fresco, discovered at Medum, dates back to a time at least four thousand years before the Christian era, and is probably the oldest picture in the world.

It is a representation of six geese, of three different species, depicted with marvellous fidelity, and a thorough appreciation of form and colouring.

Among the most distinguished in appearance and carriage of the handsome exotic species is the Magellanic goose, one of the five or six species of the Antarctic genus Chloephaga, found in Patagonia and the Magellan Islands. One peculiarity of this bird is that the sexes differ in colouring, the male being white, with grey mottlings, whereas the prevailing colour of the female is a ruddy brown,--a fine rich colour set off with some white, grey, intense cinnamon, and beautiful black mottlings.

Seen on the wing the flock presents a somewhat singular appearance, as of two distinct species associating together, as we may see when by chance gulls and rooks, or sheldrakes and black scoters, mix in one flock.

This fine bird has long been introduced into this country, and as it breeds freely it promises to become quite common. I can see it any day; but these exiles, pinioned and imprisoned in parks, are not quite like the Magellanic geese I was intimate with in former years, in Patagonia and in the southern pampas of Buenos Ayres, where they wintered every year in incredible numbers, and were called "bustards" by the natives. To see them again, as I have seen them, by day and all day long in their thousands, and to listen again by night to their wild cries, I would willingly give up, in exchange, all the invitations to dine which I shall receive, all the novels I shall read, all the plays I shall witness, in the next three years; and some other miserable pleasures might be thrown in.

Listening to the birds when, during migration, on a still frosty night, they flew low, following the course of some river, flock succeeding flock all night long; or heard from a herdsman's hut on the pampas, when thousands of the birds had encamped for the night on the plain hard by, the effect of their many voices (like that of their appearance when seen flying) was singular, as well as beautiful, on account of the striking contrasts in the various sounds they uttered. On clear frosty nights they are most loquacious, and their voices may be heard by the hour, rising and falling, now few, and now many taking part in the endless confabulation--a talkee-talkee and concert in one; a chatter as of many magpies; the solemn deep, *honk-honk*, the long, grave note changing to a shuddering sound; and, most wonderful, the fine silvery whistle of the male, steady or tremulous, now long and now short, modulated a hundred ways--wilder and more beautiful than the night-cry of the widgeon, brighter than the voice of any shore bird, or any warbler, thrush or wren, or the sound of any wind instrument.

It is probable that those who have never known the Magellanic goose in a state of nature are best able to appreciate its fine qualities in its present semi-domestic state in England. At all events the enthusiasm with which a Londoner spoke of this bird in my presence some time ago came to me rather as a surprise. It was at the studio in St John's Wood of our greatest animal painter, one Sunday evening, and the talk was partly about birds, when an elderly gentleman said that he was pleased to meet some one who would be able to tell him the name of a wonderful bird he had lately seen in St James's Park. His description was vague; he could not say what its colour was, nor what sort of beak it had, nor whether its feet were webbed or not; but it was a large tall bird, and there were two of them. It was the way this bird had comported itself towards him that had so taken him.

As he went through the park at the side of the enclosure, he caught sight of the pair some distance away on the grass, and the birds, observing that he had stopped in his walk to regard them, left off feeding, or whatever they were doing, and came to him. Not to be fed --it was impossible to believe that they had any such motive; it was solely and purely a friendly feeling towards him which caused them immediately to respond to his look, and to approach him, to salute him, in their way.

And when they had approached to within three or four yards of where he stood, advancing with a quiet dignity, and had then uttered a few soft low sounds, accompanied with certain graceful gestures, they turned and left him; but not abruptly, with their backs towards him--oh, no, they did nothing so common; they were not like other birds--they were perfect in everything; and, moving from him, half paused at intervals, half turning first to one side then the other, inclining their heads as they went. Here our old friend rose and paced up and down the floor, bowing to this side and that and making other suitable gestures, to try to give us some faint idea of the birds' gentle courtesy and exquisite grace. It was, he assured us, most astonishing; the birds' gestures and motions were those of a human being, but in their perfection immeasurably superior to anything of the kind to be seen in any Court in Europe or the world.

The birds he had described, I told him, were no doubt Upland Geese.

"Geese!" he exclaimed, in a tone of surprise, and disgust. "Are you speaking seriously?

Geese! Oh, no, nothing like geese--a sort of ostrich!"

It was plain that he had no accurate knowledge of birds; if he had caught sight of a kingfisher or green woodpecker, he would probably have described it as a sort of peacock. Of the goose, he only knew that it is a ridiculous, awkward creature, proverbial for its stupidity, although very good to eat; and it wounded him to find that any one could think so meanly of his intelligence and taste as to imagine him capable of greatly admiring any bird called a goose, or any bird in any way related to a goose.

I will now leave the subject of the beautiful antarctic goose, the "bustard" of the horsemen of the pampas, and "sort of ostrich" of our Londoner, to relate a memory of my early years, and of how I first became an admirer of the familiar domestic goose. Never since have I looked on it in such favourable conditions.

Two miles from my home there stood an old mud-built house, thatched with rushes, and shaded by a few ancient half-dead trees. Here lived a very old woman with her two unmarried daughters, both withered and grey as their mother; indeed, in appearance, they were three amiable sister witches, all very very old. The high ground on which the house stood sloped down to an extensive reed-and rush-grown marsh, the source of an important stream; it was a paradise of wild fowl, swan, roseate spoonbill, herons white and herons grey, ducks of half a dozen species, snipe and painted snipe, and stilt, plover and godwit; the glossy ibis, and the great crested blue ibis with a powerful voice. All these interested, I might say fascinated, me less than the tame geese that spent most of their time in or on the borders of the marsh in the company of the wild birds. The three old women were so fond of their geese that they would not part with one for love or money; the most they would ever do would be to present an egg, in the laying season, to some visitor as a special mark of esteem.

It was a grand spectacle, when the entire flock, numbering upwards of a thousand, stood up on the marsh and raised their necks on a person's approach. It was grand to hear them, too, when, as often happened, they all burst out in a great screaming concert. I can hear that mighty uproar now!

With regard to the character of the sound: we have seen in a former chapter that the poet Cowper thought not meanly of the domestic grey goose as a vocalist, when heard on a common or even in a farmyard. But there is a vast difference in the effect produced on the mind when the sound is heard amid its natural surroundings in silent desert places. Even hearing them as I did, from a distance, on that great marsh, where they existed almost in a state of nature, the sound was not comparable to that of the perfectly wild bird in his native haunts. The cry of the wild grey-lag was described by Robert Gray in his *Birds of the West of Scotland*. Of the bird's voice he writes: "My most recent experiences (August 1870) in the Outer Hebrides remind me of a curious effect which I noted in connection with the call-note of this bird in these quiet solitudes. I had reached South Uist, and taken up my quarters under the hospitable roof of Mr Birnie, at Grogarry ... and in the stillness of the Sabbath morning following my arrival was aroused from sleep by the cries of the grey-lags as they flew past the house. Their voices, softened by distance, sounded not unpleasantly, reminding me of the clanging of church bells in the heart of a large town."

It is a fact, I think, that to many minds the mere wildness represented by the voice of a great wild bird in his lonely haunts is so grateful, that the sound itself, whatever its quality may be, delights, and is more than the most beautiful music. A certain distinguished man of letters and Church dignitary was once asked, a friend tells me, why he lived away from society, buried in the loneliest village on the dreary East coast; at that spot where, standing on the flat desolate shore you look over the North Sea, and have no land between you and far Spitzbergen. He answered, that he made his home there because it was the only spot in England in which, sitting in his own room, he could listen to the cry of the pink-footed goose. Only those who have lost their souls will fail to understand.

The geese I have described, belonging to the three old women, could fly remarkably well, and eventually some of them, during their flights down stream, discovered at a distance of about eight miles from home the immense, low, marshy plain bordering the sea-like Plata River. There were no houses and no people in that endless green, wet land, and they liked it so well that they visited it more and more often, in small flocks of a dozen to twenty birds, going and coming all day long, until all knew the road. It was observed that when a man on foot or on horseback appeared in sight of one of these flocks, the birds at this distance from home were as wary as really wild birds, and watched the stranger's approach in alarm, and when he was still at a considerable distance rose and flew away beyond sight.

The old dames grieved at this wandering spirit in their beloved birds, and became more and more anxious for their safety. But by this time the aged mother was fading visibly into the tomb, though so slowly that long months went by while she lay on her bed, a weird-looking object--I remember her well--leaner, greyer, more ghost-like, than the silent, lean, grey heron on the marsh hard by. And at last she faded out of life, aged, it was said by her descendants, a hundred and ten years; and, after she was dead, it was found that of that great company of noble birds there remained only a small remnant of about forty, and these were probably incapable of sustained flight. The others returned no more; but whether they met their death from duck and swan shooters in the marshes, or had followed the great river down to the sea, forgetting their home, was never known. For about a year after they had ceased going back, small flocks were occasionally seen in the marshes, very wild and strong on the wing, but even these, too, vanished at last.

It is probable that, but for powder and shot, the domestic goose of Europe, by occasionally taking to a feral life in thinly-settled countries, would ere this have become widely distributed over the earth.

And one wonders if in the long centuries running to thousands of years, of tame flightless existence, the strongest impulse of the wild migrant has been wholly extinguished in the domestic goose? We regard him as a comparatively unchangeable species, and it is probable that the unexercised faculty is not dead but sleeping, and would wake again in favourable circumstances. The strength of the wild bird's passion has been aptly described by Miss Dora Sigerson in her little poem, "The Flight of the Wild Geese." The poem, oddly enough, is not about geese but about men--wild Irishmen who were called Wild Geese; but the bird's powerful impulse and homing faculty are employed as an illustration, and admirably described:--

Flinging the salt from their wings, and despair from their hearts
They arise on the breast of the storm with a cry and are gone.
When will you come home, wild geese, in your thousand strong?...
Not the fierce wind can stay your return or tumultuous sea,...
Only death in his reaping could make you return no more.

Now arctic and antarctic geese are alike in this their devotion to their distant breeding-ground, the cradle and true home of the species or race; and I will conclude this chapter with an incident related to me many years ago by a brother who was sheep-farming in a wild and lonely district on the southern frontier of Buenos Ayres. Immense numbers of upland geese in great flocks used to spend the cold months on the plains where he had his lonely hut; and one morning in August in the early spring of that southern country, some days after all the flocks had taken their departure to the south, he was out riding, and saw at a distance before him on the plain a pair of geese. They were male and female--a white and a brown bird.

Their movements attracted his attention and he rode to them. The female was walking steadily on in a southerly direction, while the male, greatly excited, and calling loudly from time to time, walked at a distance ahead, and constantly turned back to see and call to his mate, and at intervals of a few minutes he would rise up and fly, screaming, to a distance of some hundreds of yards; then finding that he had not been followed, he would return and alight at a distance of forty or fifty yards in advance of the other bird, and begin walking on as before. The female had one wing broken, and, unable to fly, had set out on her long journey to the Magellanic Islands on her feet; and her mate, though called to by that mysterious imperative voice in his breast, yet would not forsake her; but flying a little distance to show her the way, and returning again and again, and calling to her with his wildest and most piercing cries, urged her still to spread her wings and fly with him to their distant home.

And in that sad, anxious way they would journey on to the inevitable end, when a pair or family of carrion eagles would spy them from a great distance--the two travellers left far behind by

their fellows, one flying, the other walking; and the first would be left to continue the journey alone.

Since this appreciation was written a good many years ago I have seen much of geese, or, as it might be put, have continued my relations with them and have written about them too in my *Adventures among Birds* (1913). In recent years it has become a custom of mine to frequent Wells-next-the-Sea in October and November just to welcome the wild geese that come in numbers annually to winter at that favoured spot. Among the incidents related in that last book of mine about the wild geese, there were two or three about the bird's noble and dignified bearing and its extraordinary intelligence, and I wish here to return to that subject just to tell yet one more goose story: only in this instance it was about the domestic bird.

It happened that among the numerous letters I received from readers of *Birds and Man* on its first appearance there was one which particularly interested me, from an old gentleman, a retired

schoolmaster in the cathedral city of Wells. He was a delightful letter-writer, but by-and-bye our correspondence ceased and I heard no more of him for three or four years. Then I was at Wells, spending a few days looking up and inquiring after old friends in the place, and remembering my pleasant letter-writer I went to call on him. During our conversation he told me that the chapter which had impressed him most in my book was the one on the goose, especially all that related to the lofty dignified bearing of the bird, its independent spirit and fearlessness of its human masters, in which it differs so greatly from all other domestic birds. He knew it well; he had been feelingly persuaded of that proud spirit in the bird, and had greatly desired to tell me of an adventure he had met with, but the incident reflected so unfavourably on himself, as a humane and fair-minded or sportsmanlike person, that he had refrained. However, now that I had come to see him he would

make a clean breast of it.

It happened that in January some winters ago, there was a very great fall of snow in England, especially in the south and west. The snow fell without intermission all day and all night, and on the following morning Wells appeared half buried in it. He was then living with a daughter who kept house for him in a cottage standing in its own grounds on the outskirts of the town. On attempting to leave the house he found they were shut in by the snow, which had banked itself against the walls to the height of the eaves. Half an hour's vigorous spade work enabled him to get out from the kitchen door into the open, and the sun in a blue sky shining on a dazzling white and silent world. But no milkman was going his rounds, and there would be no baker nor butcher nor any other tradesman to call for orders. And there were no provisions in the house! But the milk for breakfast was the first thing needed, and so with a jug in his hand he went bravely out to try and make his way to the milk shop which was not far off.

A wall and hedge bounded his front garden on one side, and this was now entirely covered by an immense snowdrift, sloping up to a height of about seven feet. It was only when he paused to look at this vast snow heap in his garden that he caught sight of a goose, a very big snow-white bird without a grey spot in its plumage, standing within a few yards of him, about four feet from the ground.

Its entire snowy whiteness with snow for a background had prevented him from seeing it until he looked directly at it. He stood still gazing in astonishment and admiration at this noble bird, standing so motionless with its head raised high that it was like the figure of a goose carved out of some crystalline white stone and set up at that spot on the glittering snowdrift. But it was no statue; it had living eyes which without the least turning of the head watched him and every motion he made. Then all at once the thought came into his head that here was something, very good succulent food in fact, sent, he almost thought providentially, to provision his house; for how easy it would be for him as he passed the bird to throw himself suddenly upon and capture it! It had belonged to some one, no doubt, but that great snowstorm and the furious north-east wind had blown it far far from its native place and it was lost to its owner for ever. Practically it was now a wild bird free for him to take without any qualms and to nourish himself on its flesh while the snow siege lasted. Standing there, jug in hand, he thought it out, and then took a few steps towards the bird in order to see if there was any sign of suspicion in it; but there was none, only he could see that the goose without turning its head was all the time regarding him out of the corner of one eye. Finally he came to the conclusion that his best plan was to go for the milk and on his return to set the jug down by the gate when coming in, then to walk in a careless, unconcerned manner towards the door, taking no notice of the goose until he got abreast of it, and then turn suddenly and hurl himself upon it. Nothing could be easier; so away he went and in about twenty minutes was back again with the milk, to find the bird in the same place standing as before motionless in the same attitude. It was not disturbed at his coming in at the gate, nor did it show the slightest disposition to move when he walked towards it in his studied careless manner. Then, when within three yards of it, came the supreme moment, and wheeling suddenly round he hurled himself with violence upon his victim, throwing out his arms to capture it, and so great was the impulse he had given himself that he was buried to the ankles in the drift. But before going into it, in that brief moment, the fraction of a second, he saw what happened; just as his hands were about to touch it the wings opened and the bird was lifted from its stand and out of his reach as if by a miracle.

In the drift he was like a drowning man, swallowing snow into his lungs for water. For a few dreadful moments he thought it was all over with him; then he succeeded in struggling out and stood trembling and gasping and choking, blinded with snow. By-and-bye he recovered and had a look round, and lo! there stood his goose on the summit of the snow bank about three yards from the spot where it had been! It was standing as before, perfectly motionless, its long neck and head raised, and was still in appearance the snow-white figure of a carved bird, only it was more conspicuous and impressive now, being outlined against the blue sky, and as before it was regarding him out of the corner of one eye. He had never, he said, felt so ashamed of himself in his life! If the bird had screamed and fled from him it would not have been so bad, but there it had chosen to remain, as if despising his attempt at harming it too much even to feel resentment. A most uncanny bird!

it seemed to him that it had divined his intention from the first and had been prepared for his every movement; and now it appeared to him to be saying mentally: "Have you got no more plans to capture me in your clever brain, or have you quite given it up?"

Yes, he had quite, quite given it up!

And then the goose, seeing there were no more plans, quietly unfolded its wings and rose from the snowdrift and flew away over the town and the cathedral away on the further side, and towards the snow-covered Mendips; he standing there watching it until it was lost to sight in the pale sky.

CHAPTER XII

THE DARTFORD WARBLER

HOW TO SAVE OUR RARE BIRDS

The most interesting chapter in John Burroughs' *Fresh Fields* contains an account of an anxious hurried search after a nightingale in song, at a time of the year when that "creature of ebullient heart" somewhat suddenly drops into silence. A few days were spent by the author in rushing about the country in Surrey and Hampshire, with the result that once or twice a few musical throbs of sound, a trill, a short detached phrase, were heard--just enough to convince the eager listener that here was a vocalist beautiful beyond all others, and that he had missed its music by appearing a very few days too late on the scene.

During the last seven or eight years I have read this chapter several times with undiminished interest, and with a feeling of keen sympathy for the writer in his disappointment; for it is the case that I, too, all this time, have been in chase of a small British songster--a rare elusive bird, hard to find at any time as it is to hear a nightingale pour out its full song in the last week in June.

In these years I have, at every opportunity, in spring, summer, and autumn, sought for the bird in the southern half of England, chiefly in the south and south-western counties. In the Midlands, and in Devonshire, where he was formerly well known, but where the authorities say he is now extinct, I failed to find him. I found him altogether in four counties, in a few widely-separated localities; in every case in such small numbers that I was reluctantly forced to give up a long-cherished hope that this species might yet recover from the low state, with regard to numbers, in which it fingers, and be permanently preserved as a member of the British avifauna.

It would indeed hardly be reasonable to entertain such a hope, when we consider that the furze wren, or Dartford warbler, as it is named in books, is a small, frail, insectivorous species, a feeble flyer that must brave the winters at home; that down to within thirty years ago it was fairly common, though local, in the south of England, and ranged as far north as the borders of Yorkshire, and that in this period it has fallen to its present state, when but a few pairs and small colonies, wide apart, exist in isolated patches of furze in four or five, possibly six, counties.

There can be no doubt that the decline of this species, which, on account of its furze-loving habits, must always be restricted to limited areas, is directly attributable to the greed of private collectors, who are all bound to have specimens--as

many as they can get--both of the bird and its nest and eggs. Its strictly local distribution made its destruction a comparatively easy task. In 1873

Gould wrote in his large work on *British Birds*: "All the commons south of London, from Blackheath and Wimbledon to the coast, were formerly tenanted by this little bird; but the increase in the number of collectors has, I fear, greatly thinned them in all the districts near the metropolis; it is still, however, very abundant in many parts of Surrey and Hampshire." It did not long continue "very abundant." Gould was shown the bird, and supplied with specimens, by a man named Smithers, a bird-stuffer of Churt, who was at that time collecting Dartford warblers and their eggs for the trade and many private persons, on the open heath and gorse-grown country that lies between Farnham and Haslemere. Gould in the work quoted, adds: "As most British collectors must now be supplied with the eggs of the furze wren, I trust Mr Smithers will be more sparing in the future."

So little sparing was he, that when he died, but few birds were left for others of his detestable trade who came after him.

Three or four years ago I got in conversation with a heath-cutter on Milford Common, a singular and brutal-looking fellow, of the half-Gypsy Devil's Punch-Bowl type, described so ably by Baring-Gould in his *Broom Squire*. He told me that when he was a boy, about thirty-five years ago, the furze wren was common in all that part of the country, until Smithers' offer of a shilling for every clutch of eggs, had set the boys from all the villages in the district hunting for the nests. Many a shilling had he been paid for the nests he found, but in a few years the birds became rare; and he added that he had not now seen one for a very long time.

In Clark's Kennedy's *Birds of Berkshire and Buckinghamshire* we get a glimpse of the furze wren collecting business at an earlier date and nearer the metropolis. In 1868 he wrote:--"The only locality in the two counties in which this species is at all numerous, is a common in the vicinity of Sunninghill, where it is found breeding every summer, and from whence a person in the neighbourhood obtains specimens at all times of the year, with which to supply the London bird-stuffers."

When the district worked by Smithers, and the neighbouring commons round Godalming, where Newman in his *Letters of Rusticus* says he had seen the "tops of the furze quite alive with these birds," had been depleted, other favourite haunts of the little doomed furze-lover were visited, and for a time yielded a rich harvest. In a few years the bird was practically extirpated; in the sixties and seventies it was common, now there are many young ornithologists with us who have never seen it (in this country at all events) in a state of nature. In some cases even persons interested in bird life, some of them naturalists too, did not know what was going on in their immediate neighbourhood until after the bird was gone. I met with a case of the kind, a very strange case indeed, in the summer of 1899, at a place near the south coast where the bird was common after it had been destroyed in Surrey, but does not now exist. In my search for information I paid a visit to the octogenarian vicar of a small rustic village. He was a native of the parish, and loved his home above all places, even as White loved Selborne, and had been a clergyman in it for over sixty years; moreover he was, I was told, a keen naturalist, and though not a collector nor a writer of books, he knew every plant and every wild animal to be found in the parish. He better than another, I imagined, would be able to give me some authentic local information.

I found him in his study--a tall, handsome, white-haired old man, very feeble; he rose, and supporting his steps with a long staff, led me out into the grounds and talked about nature. But his memory, like his strength, was failing; he seemed, indeed, but the ruin of a man, although still of a very noble presence. What he called the vicarage gardens, where we strolled about among the trees, was a place without walks, all overgrown with grass and wildings; for roses and dahlias he showed me fennel, goat's-beard, henbane, and common hound's tongue; and when speaking of their nature he stroked their leaves and stems caressingly. He loved these better than the gardener's blooms, and so did I; but I wanted to hear about the vanished birds of the district, particularly the furze wren, which had survived all the others that were gone.

His dim eyes brightened for a moment with old pleasant memories of days spent in observing these birds; and leading me to a spot among the trees, from which there was a view of the open country beyond, he pointed to a great green down, a couple of miles away, and told me that on the other side I would come on a large patch of furze, and that by sitting quietly there for half an hour or so I might see a dozen furze wrens. Then he added: "A dozen, did I say? Why, I saw not fewer than forty or fifty flitting about the bushes the very last time I went there, and I daresay if you are patient enough you will see quite as many."

I assured him that there were no furze wrens at the spot he had indicated, nor anywhere in that neighbourhood, and I ventured to add that he must be telling me of what he had witnessed a good many years ago. "No, not so many," he returned, "and I am astonished and grieved to hear that the birds are gone--four or five years, perhaps. No, it was longer ago. You are right--I think it must be at least fifteen years since I went to that spot the last time. I am not so strong as I was, and for some years have not been able to take any long walks."

Fifteen years may seem but a short space of time to a man verging on ninety; in the mournful story of the extermination of rare and beautiful British birds for the cabinet it is in reality a long period.

Fifteen years ago the honey buzzard was a breeding species in England, and had doubtless been so for thousands of years. When the price of a "British-killed"

specimen rose to PS25, and of a "British-taken"

egg to two or three or four pounds, the bird quickly ceased to exist. Probably there is not a local ornithologist in all the land who could not say of some species that bred annually, within the limits of his own country, that it has not been extirpated within the last fifteen years.

In the instance just related, when the aged vicar, sorrying at the loss of the birds, began to recall the rare pleasure it had given him to watch them disporting themselves among the furze-bushes, something of the illusion which had been in his mind imparted itself to mine, for I could see what he was mentally seeing, and the fifteen years dwindled to a very brief space of time. Like Burroughs with the nightingale, I, too, had arrived a few days too late on the scene; the "cursed collector" had been beforehand with me, as had indeed been the case on so many previous occasions with regard to other species.

A short time after my interview with the aged vicar, at an inn a very few miles from the village, I met a person who interested me in an exceedingly unpleasant way. He was a big repulsive-looking man in a black greasy coat--a human animal to be avoided; but I overheard him say something about rare birds which caused me to put on a friendly air and join in the talk. He was a Kentish man who spent most of his time in driving about from village to village, and from farm to farm, in the southern counties, in search of bargains, and was prepared to buy for cash down anything he could find cheap, from an old teapot, or a print, or copper scuttle, to a horse, or cart, or pig, or a houseful of furniture. He also bought rare birds in the flesh, or stuffed, and was no doubt in league with a good many honest gamekeepers in those counties. I had heard of "travellers" sent out by the great bird stuffers to go the rounds of all the big estates in some parts of England, but this scoundrel appeared to be a traveller in the business on his own account. I asked him if he had done anything lately in Dartford warblers.

He at once became confidential, and said he had done nothing but hoped shortly to do something very good indeed. The bird, he said, was supposed to be extinct in Kent, and on that account specimens obtained in that county would command a high price. Now he had but recently discovered that a few--two or three pairs--existed at one spot, and he was anxious to finish the business he had on hand so as to go there and secure them. In answer to further questions, he said that the birds were in a place where they could not very well be shot, but that made no difference; he had a simple, effective way of getting them without a gun, and he was sure that not one would escape him.

On my mentioning the fact that the Kent County Council had obtained an order for an all the year round protection of this very bird, he looked at me out of the corners of his eyes and laughed, but said nothing. He took it as a rather good joke on my part.

There is not the slightest doubt that our wealthy private collectors have created the class of injurious wretches to which this man belonged.

*

To some who have glanced at a little dusty, out of shape mummy of a bird, labelled "Dartford Warbler," in a museum, or private collection, or under a glass shade, it may seem that I speak too warmly of the pleasure which the sight of the small furze-lover can give us. They have never seen it in a state of nature, and probably never will. When I consider all these British Passeres, which, seen at their best, give most delight to the aesthetic sense--the jay, the "British Bird of Paradise," as I have ventured to call it, displaying his vari-coloured feathers at a spring-time gathering; the yellow-green, long-winged wood wren, most aerial and delicate of the woodland warblers; the kingfisher, flashing turquoise blue as he speeds by; the elegant fawn-coloured, black-bearded tit, clinging to the grey-green, swaying reeds, and springing from them with a bell-like note; and the rose-tinted narrow-shaped bottle-tit as he drifts by overhead in a flock; the bright, lively goldfinch scattering the silvery thistle-down on the air; the crossbill, that quaint little many-coloured parrot of the north, feeding on a pine-cone; the grey wagtail exhibiting his graceful motions; and the golden-crested wren, seen suspended motionless with swiftly vibrating wings above his mate concealed among the clustering leaves, in appearance a great green hawk-moth, his opened and flattened crest a shining, flame-coloured disc or shield on his head,--when I consider all these, and others, I find that the peculiar charm of each does not exceed in degree that of the furze wren--seen at *his* best. He is of the type of the white-throat, but idealised; the familiar brown, excitable Sylvia, pretty as he is and welcome to our hedges in April, is in appearance but a rough study for the smaller, more delicately-fashioned and richly-coloured Melizophilus, or furze-lover. On account of his excessive rarity he can now be seen at his best only by those who are able to spend many days in searching and in watching, who have the patience to sit motionless by the hour; and at length the little hideling, tired of concealment or overcome by curiosity, shows himself and comes nearer and nearer, until the ruby red of the small gem-like eye may been seen without aid to the vision. A sprite-like bird in his slender exquisite shape and his beautiful fits of excitement; fantastic in his motions as he flits and flies from spray to spray, now hovering motionless in the air like the wooing gold-crest, anon dropping on a perch, to sit jerking his long tail, his crest raised, his throat swollen, chiding when he sings and singing when he chides, like a refined and lesser sedge warbler in a frenzy, his slate-black and chestnut-red plumage showing rich and dark against the pure luminous yellow of the massed furze blossoms. It is a sight of fairy-like bird life and of flower which cannot soon be forgotten. And I do not think that any man who has in him any love of nature and of the beautiful can see such a thing, and exist with its image in his mind, and not regard with an extreme bitterness of hatred those among us whose particular craze it is to "collect" such creatures, thereby depriving us and our posterity of the delight the sight of them affords.

Of many curious experiences I have met in my quest of the rare little bird, or of information concerning it, I have related two or three: I have one more to give--assuredly the strangest of all. I was out for a day's ramble with the members of a Natural History Society, at a place the name of which must not be told, and was walking in advance of the others with a Mr A., the leading ornithologist of the county, one whose name is honourably known to all naturalists in the kingdom. The Dartford warbler, he said in the course of conversation, had unhappily long been extinct in the county. Now it happened that among those just behind us there was another local naturalist, also well known outside his own county--Mr B., let us call him. When I separated from my companion this gentleman came to my side, and said that he had overheard some of our talk, and he wished me to know that Mr A. was in error in saying that the Dartford warbler was extinct in the county. There was one small colony of three or four pairs to be found at a spot ten to eleven miles from where we then were; and he would be glad to take me to the place and show me the birds. The existence of this small remnant had been known for several years to half a dozen persons, who had jealously kept the secret;--to their great regret they had had to keep it from their best friend and chief supporter of their Society, Mr A., simply because it would not be safe with him. He was enthusiastic about the native bird life, the number of species the county could boast, etc., and sooner or later he would incautiously speak about the Dartford warbler, and the wealthy local collectors would hear of it, with the result that the birds would quickly be gathered into their cabinets.

My informant went on to say that the greatest offenders were four or five gentlemen in the place who were zealous collectors. The county had obtained a stringent order, with all-the-year-round protection for its rare species. Much, too, had been done by individuals to create a public opinion favourable to bird protection, and among the educated classes there was now a strong feeling against the destruction by private collectors of all that was best worth preserving in the local wild bird life. But so far not the slightest effect had been produced in the principal offenders. They would have the rare birds, both the resident species and the occasional visitants, and paid liberally for all specimens. Bird-stuffers, gamekeepers--their own and their neighbours'--fowlers, and all those who had a keen eye for a feathered rarity, were in their pay; and so the destruction went merrily on. The worst of it was that the authors of the evil, who were not only law-breakers themselves, but were paying others to break the law, could not be touched; no one could prosecute nor openly denounce them because of their important social position in the county.

There was nothing new to me in all this: it was an old familiar story; I have given it fully, simply because it is an accurate statement of what is being done all over the country. There is not a county in the kingdom where you may not hear of important members of the community who are collectors of birds and their eggs, and law-breakers, both directly and indirectly, every day of their lives. They all take, and pay for, every rare visitant that comes in their way, and also require an unlimited supply of the rarer resident species for the purpose of exchange with other private collectors in distant counties. In this way our finest species are gradually being extirpated. Within the last few years we have seen the disappearance (as breeding species) of the ruff and reeve, marsh harrier, and honey buzzard; and the species now on the verge of extinction, which will soon follow these and others that have gone before, if indeed some of them have not already gone, are the sea-eagle, osprey, kite, hen harrier, Montagu's harrier, stone curlew, Kentish plover, dotterel, red-necked phalarope, roseate tern, bearded tit, grey-lag goose, and great skua. These in their turn will be followed by the chough, hobby, great black-backed gull, furze wren, crested tit, and others.

These are the species which, as things are going, will absolutely and for ever disappear, as residents and breeders, from off the British Islands. Meanwhile other species that, although comparatively rare, are less local in their distribution, are being annually exterminated in some parts of the country: it is poor comfort to the bird lover in southern England to know that many species that formerly gave life and interest to the scene, and have lately been done to death there, may still be met with in the wilder districts of Scotland, or in some forest in the north of Wales. Finally, we have among our annual visitants a considerable number of species which have either bred in these islands in past times (some quite recently), or else would probably remain to breed if they were not immediately killed on arrival--bittern, little bittern, night heron, spoonbill, stork, avocet, black tern, hoopoe, golden oriole, and many others of less well-known names.

This is the case, and that it is a bad one, and well-nigh hopeless, no man will deny. Nevertheless, I believe that it may be possible to find a remedy.

That "destruction of beautiful things," about which Ruskin wrote despairingly, "of late ending in perfect blackness of catastrophe, and ruin of all grace and glory in the land," has fallen, and continues to fall, most heavily on the beautiful bird life of our country. But the destruction has not been unremarked and unlamented, and the existence of a strong and widespread public feeling in favour of the preservation of our wild birds has of late shown itself in many ways, especially in the unopposed legislation on the subject during the last few years, and the willingness that Government and Parliament have shown recently to consider a new Act.

There is no doubt that this feeling will grow until it becomes too strong even for the selfish Philistines, who are blind to all grace and glory in nature, and incapable of seeing anything in a rare and beautiful bird but an object to be collected.

Those who in the years to come will inherit the numberless useless private collections now being formed will make haste to rid themselves of such unhappy legacies, by thrusting them upon local museums, or by destroying them outright in their anxiety to have it forgotten that one of their name had a part in the detestable business of depriving the land of these wonderful and beautiful forms of life--a life which future generations would have cherished as a dear and sacred possession.

But we cannot afford to wait: we have been made too poor in species already, and are losing something further every year; we want a remedy now.

So far two suggestions have been made. One is an alteration in the existing law, which will allow the infliction of far heavier fines on offenders. All those who are acquainted with collectors and their ways will at once agree that increased penalties will not meet the case; that the only effect of such an alteration in the law would be to make collectors and the persons employed by them more careful than they have yet found it necessary to be. The other suggestion vaguely put forth is that something of the nature of a private inquiry agency should be established to find out the offenders, and that they should be pilloried in the columns of some widely-circulating journal, a method which has been tried with some success in the cases of other classes of obnoxious persons. This suggestion may be dismissed at once as of no value; not one offence in a hundred would be discovered by such means, and the greatest sinners, who are not infrequently the most intelligent men, would escape scot free.

Perhaps I should have said that *three* suggestions have been made, for there is yet another, put forward by Mr Richard Kearton in one of his late books.

He is thoroughly convinced, he tells us, that the County Council orders are perfectly useless in the case of any and every rare bird which collectors covet; and on that point we are all agreed; he then says: "We should select a dozen species admitted by a committee of practical ornithologists to be in danger, and afford them personal protection during the whole of the breeding season by placing reliable watchers, night and day, upon the nesting-ground."

Watchers provided and paid by individuals and associations have been in existence these many years, and this is undoubtedly the best plan in the case of all species which breed in colonies. These are mostly sea-birds--gulls, terns, cormorants, guillemots, razor-bills, *etc.* Our rare birds are distributed over the country, and in the case of some, if a hundred pairs of a species exist in the British Islands, a hundred or two hundred watchers would have to be engaged. But who that has any knowledge of what goes on in the collecting world does not know that the guarded birds would be the first to vanish? I have seen such things--pairs of rare birds breeding in private grounds, where the keepers had strict orders to watch over them, and no stranger could enter without being challenged, and in a little while they have mysteriously disappeared. The "watcher" is good enough on the exposed sea-coast or island where an eye is kept on his doings, and where the large number of birds in his charge enables him to do a little profitable stealing and still keep up an appearance of honesty. I have visited most of the watched colonies, and therefore know. The watchers, who were paid a pound a week for guarding the nests, were not chary of their hints, and I have also been told in very plain words that I could have any eggs I wanted.

It is hardly necessary to say here that the proposed alteration in the law to make it protective of all species will, so far as the private collector is concerned, leave matters just as they are.

There is really only one way out of the difficulty,--one remedy for an evil which grows in spite of penalties and of public opinion,--namely, a law to forbid the making of collections of British birds by private persons. If all that has been done in and out of Parliament since 1868 to preserve our wild birds--not merely the common abundant species, which are not regarded by collectors, but *all* species--is not to be so much labour wasted, such a law must sooner or later be made. It will not be denied by any private collector, whether he clings to the old delusion that it is to the advantage of science that he should have cabinets full of "British killed"

specimens or not,--it will not be denied that the drain on our wild bird life caused by collecting is a constantly increasing one, and that no fresh legislation on the lines of previous bird protection Acts can arrest or diminish that drain. Thirty years ago, when the first Act was passed, which prohibited the slaughter of sea-birds during the breeding season, the drain on the bird life which is valued by collectors was far less than it is now; not only because there are a dozen or more collectors now where there was one in the sixties, but also because the business of collecting has been developed and brought to perfection. All the localities in which the rare resident species may be looked for are known, while the collectors all over the country are in touch with each other, and have a system of exchanges as complete as it is deadly to the birds. Then there is the money element; bird-collecting is not only the hobby of hundreds of persons of moderate means and of moderate wealth, but, like horse-racing, yachting, and other expensive forms of sport, it now attracts the very wealthy, and is even a pastime of millionaires.

All this is a familiar fact, and clearly shows that without such a law as I have suggested it has now become impossible to save the best of our wild bird life.

The collectors will doubtless cry out that such a law would be a monstrous injustice, and an unwarrantable interference with the liberty of the subject; that there is really no more harm in collecting birds and their eggs than in collecting old prints, Guatemalan postage stamps, samplers, and first editions of minor poets; that to compel them to give up their treasures, which have cost them infinite pains and thousands of pounds to get together, and to abandon the pursuit in which their happiness is placed, would be worse than confiscation and downright tyranny; that the private collectors cannot properly be described as law-breakers and injurious persons, since they count among their numbers hundreds of country gentlemen of position, professional men (including clergymen), noblemen, magistrates, and justices of the peace, and distinguished naturalists--all honourable men.

To put in one word on this last very delicate point: Where, in collecting, does the honourable man draw the line, and sternly refuse to enrich his cabinet with a long-wished-for specimen of a rare British species?--a specimen "in the flesh," not only "British killed" but obtained in the county; not killed wantonly, nor stolen by some poaching rascal, but unhappily shot in mistake for something else by an ignorant young under-keeper, who, in fear of a wigging, took it secretly to a friend at a distance and gave it to him to get rid of. The story of the unfortunate killing of the rare bird varies in each case when it has to be told to one whose standard of morality is very high even with regard to his hobby.

My experience is, that where there are collectors who are men of means, there you find their parasites, who know how to treat them, and who feed on their enthusiasms.

In my rambles about the country during the last few years, I have neglected no opportunity of conversing with landowners and large tenants on this subject, and, with the exception of one man, all those I have spoken to agreed that owners generally--not nine in every ten, as I had put it, but ninety-nine in every hundred--would gladly welcome a law to put down the collecting of British birds by private persons. The one man who disagreed is the owner of an immense estate, and he was the bitterest of all in denouncing the scoundrels who came to steal his birds; and if a law could be made to put an end to such practices he would, he said, be delighted; but he drew the line at forbidding a man to collect birds on his own property. "No, no!" he concluded; "*that* would be an interference with the liberty of the subject." Then it came out that he was a collector himself, and was very proud of the rare species in his collection! If I had known that before, I should not have gone out of my way to discuss the subject with him.

Clearly, then, there is a very strong case for legislation. How strong the case is I am not yet able to show, my means not having enabled me to carry out an intention of discussing the subject with a much greater number of landowners, and of addressing a circular later stating the case to all the landlords and shooting-tenants in the country.

That remains to be done; in the meantime this chapter will serve to bring the subject to the attention of a considerable number of persons who would prefer that our birds should be preserved rather than that they should be exterminated in the interests of a certain number of individuals whose amusement it is to collect such objects.

That a law on the lines suggested will be made sooner or later is my belief: that it may come soon is my hope and prayer, lest we have to say of the Dartford warbler, and of twenty other species named in this chapter, as we have had to say of so many others that have gone

The beautiful is vanished and returns not.

*

NOTE.--The foregoing chapter, albeit written so many years ago, is still "up-to-date"--still represents without a shadow of a shade of difference the state of the case. The extermination of our rare birds and "occasional visitors" still goes merrily on in defiance of the law, and the worst offenders are still received with open arms by the British Ornithologists' Union. Indeed, that Society, from the point of view of many of its members would have no *raison d'etre* if membership were denied to the private collector of rare "British killed" birds and their eggs and to the "scientific" ornithologist whose mission is to add several new species annually to the British list. They still dine together and exhibit their specimens to one another. On the last occasion of my attending one of these meetings a member exhibited a small bird "in the flesh"--a bird from some far country which had been shot somewhere on the east coast and was so knocked to pieces by the shot that the ornithologists had great difficulty in identifying it. Although a collector himself he was anxious to dispose of the specimen, but none of his brother collectors would give him a five-pound note for it owing to its condition. It was handed round and examined and discussed by all the authorities present. I stood apart, looking at a group of ornithologists bending over the shattered specimen, all talking and arguing, when another member who by chance was not a collector moved to my side and whispered in my ear: "Just like a lot of little children!"

Is it not time to say to these "little children" that they must find a new toy--a fresh amusement to fill their vacant hours: that birds--living flying birds--are a part of nature, of this visible world in this island, the dwelling-place of some forty-five or fifty millions of souls; that these millions have a right in the country's wild life too--surely a better one than that of a few hundreds of gentlemen of leisure who have money to hire gamekeepers, bird-stuffers, wild-fowlers, and many others, to break the law for them, and to take the punishment when any is given?

By *saying* it will be understood that I mean enacting a law to prohibit private collection. It is surely time. But what prospects are there of such an Act being passed by a Parliament which has spent six years playing with a Plumage Prohibition Bill!

Well, just now we have a committee appointed by the Government to consider the whole question of bird protection with a view to fresh legislation. Will this committee recommend the one and only way to put a stop to the continuous destruction of our rarer birds? I don't think so. For such a law would be aimed at those of their own class, at their friends, at themselves.

At the end of the chapter I gave an account of an interview I had with a great landowner who happened to be a collector, and who cried out that such a law as the one I suggested would be an unwarrantable interference with the liberty of the subject.

Another interview years later was with one who is not only a landowner, the head of a branch of a great family in the land, but a great power in the political world as well, and, finally, (*not* wonderful to relate) a great "protector of birds." "No,"

he said warmly, "I will not for a moment encourage you to hope that any good will come of such a proposal. If any person should bring in such a measure I would do everything in my power to defeat it. I am a collector myself and I am perfectly sure that such an interference with the liberty of the subject would not be tolerated."

That, I take it, is or will be the attitude of the committee now considering the subject of our wild bird life and its better protection.

CHAPTER XIII

VERT--VERT; OR PARROT GOSSIP

I am not an admirer of pet parrots. To me, and I have made the discovery that to many others too, it is a depressing experience, on a first visit to nice people, to find that a parrot is a member of the family. As a rule he is the most important member.

When I am compelled to stand in the admiring circle, to look on and to listen while he exhibits his weary accomplishments, it is but lip service that I render: my eyes are turned inward, and a vision of a green forest comes before them resounding with the wild, glad, mad cries of flocks of wild parrots. This is done purposely, and the sound which I mentally hear and the sight of their vari-coloured plumage in the dazzling sunlight are a corrective, and keep me from hating the bird before me because of the imbecility of its owners. In his proper place, which is not in a tin cage in a room of a house, he is to be admired above most birds; and I wish I could be where he is living his wild life; that I could have again a swarm of parrots, angry at my presence, hovering above my head and deafening me with their outrageous screams. But I cannot go to those beautiful distant places--I must be content with an image and a memory of things seen and heard, and with the occasional sight of a bird, or birds, kept by some intelligent person; also with an occasional visit to the Parrot House in Regent's Park. There the uproar, when it is at its greatest, when innumerable discordant voices, shrill and raucous, unite in one voice and one great cry, and persons of weak nerves stop up their ears and fly from such a pandemonium, is highly exhilarating.

Of the most interesting captive parrots I have met in recent years I will speak here of two. The first was a St Vincent bird, *Chrysotis guildingi*, brought home with seven other parrots of various species by Lady Thompson, the wife of the then Administrator of the Island. This is a handsome bird, green, with blue head and yellow tail, and is a member of an American genus numbering over forty species. He received his funny specific name in compliment to a clergyman who was a zealous collector not of men's souls, but of birds' skins.

To ornithologists this parrot is interesting on account of its rarity. For the last thirty years it has existed in small numbers; and as it is confined to the island of St Vincent it is feared that it may become extinct at no distant date. Altogether there are about five hundred species of parrots in the world, or about as many parrots as there are species of birds of all kinds in Europe, from the great bustard, the hooper swan, and golden eagle, to the little bottle-tit whose minute body, stript of its feathers, may be put in a lady's thimble. And of this multitude of parrots the St Vincent Chrysotis, if it still exists, is probably the rarest.

The parrot I have spoken of, with his seven travelling companions, arrived in England in December, and a few days later their mistress witnessed a curious thing. On a cold grey morning they were enjoying themselves on their perches in a well-warmed room in London, before a large window, when suddenly they all together emitted a harsh cry of alarm or terror--the sound which they invariably utter on the appearance of a bird of prey in the sky, but at no other time. Looking up quickly she saw that snow in big flakes had begun to fall. It was the birds' first experience of such a phenomenon, but they had seen and had been taught to fear something closely resembling falling flakes --flying feathers to wit. The fear of flying feathers is universal among species that are preyed upon by hawks. In a majority of cases the birds that exhibit terror and fly into cover or sit closely have never actually seen that winged thunderbolt, the peregrine falcon, strike down a duck or pigeon, sending out a small cloud of feathers; or even a harrier or sparrow-hawk pulling out and scattering the feathers of a bird it has captured, but a tradition exists among them that the sight of flying feathers signifies danger to bird life.

When I was in the young barbarian stage, and my playmates were gaucho boys on horseback on the pampas, they taught me to catch partridges in their simple way with a slender cane twenty to twenty-five feet long, a running noose at its tip made from the fine pliant shaft of a rhea's wing feather.

The bird was not a real partridge though it looks like it, but was the common or spotted tinamou of the plains, *Nothura maculosa,* as good a table bird as our partridge. Our method was, when we flushed a bird, to follow its swift straight flight at a gallop, and mark the exact spot where it dropped to earth and vanished in the grass, then to go round the spot examining the ground until the tinamou was detected in spite of his protective colouring sitting close among the dead and fading grass and herbage. The cane was put out, the circle narrowed until the small noose was exactly over the bird's head, so that when he sprang into the air on being touched by the slender tip of the cane he caught and strangled himself. To make the bird sit tight until the noose was actually over his head, we practised various tricks, and a very common one was, on catching sight of the close-squatting partridge, to start plucking feathers from a previously-killed bird hanging to our belt and scatter them on the wind.

Sometimes we were saved the trouble of scattering feathers when we were followed by a pair of big carrion hawks on the look-out for an escaped bird or for any trifle we throw to them to keep them with us.

The effect was the same in both cases; the sight of the flying feathers was just as terrifying as that of the big hovering hawks, and caused the partridge to sit close.

This way of taking the tinamou may seem unsportsmanlike.

Well, if I were a boy in a wild

land again--with my present feelings about bird life, I mean--I should not do it. Nor would I shoot them; for I take it that the gun is the deadliest instrument our cunning brains have devised to destroy birds in spite of their bright instinct of self-preservation, their faculty of flight, and their

intelligence. It is a hundred times more effective than the boy-on-horseback's long cane with its noose made of an ostrich feather--therefore more unsportsmanlike.

To return. The resemblance of falling flakes to flying white feathers does not deceive birds accustomed to the sight of snow: it is very striking, nevertheless, and so generally recognised that most persons in Europe have heard of the old woman plucking her geese in the sky. It is curious to find the subject discussed in Herodotus. In Book IV.

he says: "The Scythians say that those lands which are situated in the northernmost parts of their territories are neither visible nor practicable by reason of the feathers that fall continually on all sides; for the earth is so entirely covered, and the air is so full of these feathers, that the sight is altogether obstructed." Further on he says: "Touching the feathers ... my opinion is that perpetual snows fall in those parts, though probably in less quantity during the summer than in winter, and whoever has observed great abundance of snow falling will easily comprehend what I say, for snow is not unlike feathers."

Probably the Scythians had but one word to designate both. To go back to the St Vincent parrot. Concerning a bird of that species I have heard, and cannot disbelieve, a remarkable story.

During the early years of the last century a gentleman went out from England to look after some landed property in the island, which had come to him by inheritance, and when out there he paid a visit to a friend who had a plantation in the interior.

His friend was away when he arrived, and he was conducted by a servant into a large, darkened, cool room; and, tired with his long ride in the hot sun, he soon fell asleep in his chair. Before long a loud noise awoke him, and from certain scrubbing sounds he made out that a couple of negro women were engaged in washing close to him, on the other side of the lowered window blinds, and that they were quarrelling over their task. Of course the poor women did not know that he was there, but he was a man of a sensitive mind and it was a torture to him to have to listen to the torrents of exceedingly bad language they discharged at one another. It made him angry. Presently his friend arrived and welcomed him with a hearty hand-shake and asked him how he liked the place. He answered that it was a very beautiful place, but he wondered how his friend could tolerate those women with their tongues so close to his windows. Women with their tongues!

What did he mean? exclaimed the other in great surprise. He meant, he said, those wretched nigger washerwomen outside the window. His host thereupon threw up the blind and both looked out: no living creature was there except a St Vincent parrot dozing on his perch in the shaded verandah. "Ah, I see, the parrot!" said his friend. And he apologised and explained that some of the niggers had taken advantage of the bird's extraordinary quickness in learning to teach him a lot of improper stuff.

Another parrot, which interested me more than the St Vincent bird, was a member of the same numerous genus, a double-fronted amazon, *Chrysotis lavalainte*, a larger bird, green with face and fore-part of head pure yellow, and some crimson colour in the wings and tail. I came upon it at an inn, the Lamb, at Hindon, a village in the South Wiltshire downs. One could plainly see that it was a very old bird, and, judging from the ragged state of its plumage, that it had long fallen into the period of irregular or imperfect moult--"the sere, the yellow leaf" in the bird's life. It also had the tremor of the very aged--man or bird. But its eyes were still as bright as polished yellow gems and full of the almost uncanny parrot intelligence. The voice, too, was loud and cheerful; its call to its mistress--"Mother, mother!" would ring through the whole rambling old house. He talked and laughed heartily and uttered a variety of powerful whistling notes as round and full and modulated as those of any grey parrot. Now, all that would not have attracted me much to the bird if I had not heard its singular history, told to me by its mistress, the landlady. She had had it in her possession fifty years, and its story was as follows:--

Her father-in-law, the landlord of the Lamb, had a beloved son who went off to sea and was seen and heard of no more for a space of fourteen years, when one day he turned up in the possession of a sailor's usual fortune, acquired in distant barbarous lands--a parrot in a cage! This he left with his parents, charging them to take the greatest care of it, as it was really a very wonderful bird, as they would soon know if they could only understand its language, and he then began to make ready to set off again, promising his mother to write this time and not to stay away more than five or at most ten years.

Meanwhile, his father, who was anxious to keep him, succeeded in bringing about a meeting between him and a girl of his acquaintance, one who, he believed, would make his son the best wife in the world. The young wanderer saw and loved, and as the feeling was returned he soon married and endowed her with all his worldly possessions, which consisted of the parrot and cage. Eventually he succeeded his father as tenant of the Lamb, where he died many years ago; the widow was grey when I first knew her and old like her parrot; and she was like the bird too in her youthful spirit and the brilliance of her eyes.

Her young sailor had picked up the bird at Vera Cruz in Mexico. He saw a girl standing in the market place with the parrot on her shoulder. She was talking and singing to the bird, and the bird was talking, whistling, and singing back to her--singing snatches of songs in Spanish. It was a wonderful bird, and he was enchanted and bought it, and brought it all the way back to England and Wiltshire. It was, the girl had told him, just five years old, and as fifty years had gone by it was, when I first knew it, or was supposed to be, fifty-five. In its Wiltshire home it continued to talk and sing in Spanish, and had two favourite songs, which delighted everybody, although no one could understand the words. By and by it took to learning words and sentences in English, and spoke less in Spanish year after year until in about ten to twelve years that language had been completely forgotten. Its memory was not as good as that of Humboldt's celebrated parrot of the Maipures, which had belonged to the Apures tribe before they were exterminated by the Caribs. Their language perished with them, only the long-living parrot went on talking it. This parrot story took the fancy of the public and was re-told in a hundred books, and was made the subject of poems in several countries--one by our own "Pleasures of Hope"

Campbell.

Nevertheless I thought it would be worth while trying a little Spanish on old Polly of the Lamb, and thought it best to begin by making friends.

It was of little use to offer her something to eat. Poll was a person who rather despised sweeties and kickshaws.

It had been the custom of the house for half a century to allow Polly to eat what she liked and when she liked, and as she--it was really a he--was of a social disposition she preferred taking her meals with the family and eating the same food. At breakfast she would come to the table and partake of bacon and fried eggs, also toast and butter and jam and marmalade, at dinner it was a cut off the joint with (usually) two vegetables, then pudding or tart with pippins and cheese to follow. Between meals she amused herself with bird seed, but preferred a meaty mutton-bone, which she would hold in one hand or foot and feed on with great satisfaction. It was not strange that when I held out food for her she took it as an insult, and when I changed my tactics and offered to scratch her head she lost her temper altogether, and when I persisted in my advances she

grew dangerous and succeeded in getting in several nips with her huge beak, which drew blood from my fingers.

It was only then, after all my best blandishments had been exhausted, and when our relations were at their worst, that I began talking to her in Spanish, in a sort of caressing falsetto like a "native" girl, calling her "Lorito" instead of Polly, coupled with all the endearing epithets commonly used by the women of the green continent in addressing their green pets. Polly instantly became attentive. She listened and listened, coming down nearer to listen better, the one eye she fixed on me shining like a fiery gem. But she spoke no word, Spanish or English, only from time to time little low inarticulate sounds came from her. It was evident after two or three days that she was powerless to recall the old lore, but to me it also appeared evident that some vague memory of a vanished time had been evoked--that she was conscious of a past and was trying to recall it. At all events the effect of the experiment was that her hostility vanished, and we became friends at once. She would come down to me, step on to my hand, climb to my shoulder, and allow me to walk about with her.

It saddened me a few months later to receive a letter from her mistress announcing Polly's death, on 2nd December 1909.

I have thought since that this bird, instead of being only five years old when bought, was probably aged twenty-five years or more. Naturally, the girl who had been sent into the market-place to dispose of the bird would tell a possible buyer that it was young; the parrots one wants to buy are generally stated to be five years old. However, it may be that the bird grew old before its time on account of its extraordinary dietary. The parrot may have an adaptive stomach, still, one is inclined to think that half a century of fried eggs and bacon, roast pork, boiled beef and carrots, steak and onions, and stewed rabbit must have put a rather heavy strain on its system.

Many parrots have lived longer than Polly in captivity, long as her life was; and here it strikes me as an odd circumstance that Polly's specific name was bestowed on the species, the double-fronted amazon, as a compliment to the distinguished French ornithologist, La Valainte, who has himself recorded the greatest age to which a captive parrot has been known to attain. This bird was the familiar African grey species. He says that it began to lose its memory at the age of sixty, to moult irregularly at sixty-five, that it became blind at ninety, and died aged ninety-three.

We may well believe that if parrots are able to exist for fifty years to a century in the unnatural conditions in which they are kept, caged or chained in houses, overfed, without using their enormously-developed wing-muscles, the constant exercise of which must be necessary to perfect health and vigour, their life in a state of nature must be twice as long.

To return to parrots in general. This bird has perhaps more points of interest for us than any other of the entire class: his long life, unique form, and brilliant colouring, extreme sociability, intelligence beyond that of most birds, and, last, his faculty of imitating human speech more perfectly than the birds of other families.

The last is to most persons the parrot's greatest distinction; to me it is his least. I do not find it so wonderful as the imitative faculty of some mocking birds or even of our delightful little marsh-warbler, described in another book. This may be because I have never had the good fortune to meet with a shining example, for we know there is an extraordinary difference in the talking powers of parrots, even in those of the same species--differences as great, in fact, as we find in the reasoning faculty between dog and dog, and in the songs of different birds of the same species. Not once but on several occasions I have heard a song from some common bird which took my breath away with astonishment.

I have described in another book certain blackbirds of genius I have encountered. And what a wonderful song that caged canary in a country inn must have had, which tempted the great Lord Peterborough, a man of some shining qualities, to get the bird from its mistress, an old woman who loved it and refused to sell it to him, by means of a dishonest and very mean trick. Denied the bird, he examined it minutely and went on his way. In due time he returned with a canary closely resembling the one he wanted in size, colour, and markings, concealed on his person. He ordered dinner, and when the good woman was gone from the room to prepare it, changed his bird for hers, then, having had his meal, went on his way rejoicing. Still he was curious to learn the effect of his trick, and whether or not she had noticed any difference in her loved bird; so, after a long interval, he came once more to the inn, and seeing the bird in its cage in the old place began to speak in praise of its beautiful singing as he had heard it and remembered it so well.

She replied sadly that since he listened to and wanted to buy it an unaccountable change had come over her bird. It was silent for a spell, perhaps sick, but when it resumed singing its voice had changed and all the beautiful notes which everyone admired were lost. The great man expressed his regret, and went away chuckling at his deliciously funny joke.

The ordinary talking parrot is no more to me than the ordinary or average canary, piping his thin expressionless notes; he is a prodigy I am pleased not to know. On the other hand there are numerous authenticated cases of parrots possessed of really surprising powers, and it was doubtless the mimicking powers of such birds of genius which suggested such fictions as that of the Tota Kuhami in the East; and in Europe, Gresset's lively tale of *Vert Vert* and the convent nuns.

It was perhaps a parrot of this rare kind which played so important a part in the early history of South America. It is nothing but a legend of the Guarani nation, which inhabit Paraguay, nevertheless I do believe that we have here an account mainly true of an important event in the early history of the race or nation. This parrot is not the impossible bird of the fictitious Tota Kahami order we all know, who not only mimics our speech but knows the meaning of the words he utters. He was nothing but a mimic, exceptionally clever, and the moral of the story is the familiar one that great events may proceed from the most trivial causes, once the passions of men are inflamed.

The tradition was related centuries ago to the Jesuit Fathers in Paraguay, and I give it as they tell it, briefly.

*

In the beginning a great canoe came over the waters from the east and was stranded on the shores of Brazil. Out of the canoe came the brothers Tupi and Guarani and their sons and daughters with their husbands and wives and their children and children's children.

Tupi was the leader, and being the eldest was called the father, and Tupi said to his brother: Behold, this great land with all its rivers and forests, abounding in fish and birds and beasts and fruit, is ours, for there are no other men dwelling in it; but we are few in number, let us therefore continue to live together with our children in one village.

Guarani consented, and for many years they lived together in peace and amity like one family, until at last there came a quarrel to divide them. And it was all about a parrot that could talk and laugh and sing just like a man. A woman first found it in the forest, and not wishing to burden herself with the rearing of it she gave it to another woman. So well did it learn to talk from its new mistress that everybody admired it and it grew to be the talk of the village.

Then the woman who had found and brought it, seeing how much it was admired and talked about, went and claimed it as her own. The other refused to give it up, saying that she had reared it and had taught it all it knew, and by doing so had become its rightful owner.

Now, no person could say which was in the right, and the dispute was not ended and tongues continued wagging until the husbands of the two women became engaged in the quarrel. And then brothers and sisters and cousins were drawn into it, until the whole village was full of bitterness and strife, all because of the parrot, and men of the same blood for the first time raised weapons against one another.

And some were wounded and others killed in open fight, and some were treacherously slain when hunting in the forest.

Now when things had come to this pass Tupi the Father, called his brother to him and said: O brother Guarani, this is a day of grief to us who had looked to the spending of our remaining years together with all our children at this place where we have lived so long. Now this can no longer be on account of the great quarrel about a parrot, and the shedding of blood; for only by separating our two families can we save them from destroying one another.

Come then, let us divide them and lead them away in opposite directions, so that when we settle again they may be far apart. Guarani consented, and he also said that Tupi was the elder and their head, and was called the Father, and it was therefore in his right to remain in possession of the village and of all that land and to end his days in it. He, on his part, would call his people together and lead them to a land so distant that the two families would never see nor hear of each other again, and there would be no more bitter words and strife between them.

Then the two old brothers bade each other an eternal farewell, and Guarani led his people south a great distance and travelled many moons until he came to the River Paraguay, and settled there; and his people still dwell there and are called by his name to this day.

Only, I beg to add, they do not call their nation by that word, as the Spanish colonists first spelt it in their carelessness, and as they pronounce it.

Heaven knows how *we* pronounce it! They, the Guarani people, call themselves Wa-ra-na-ee, in a soft musical voice. Also they call their river, which we spell Paraguay, and pronounce I don't know how, Pa-ra-wa-ee.

CHAPTER XIV

SOMETHING PRETTY IN A GLASS CASE

It was said by a Norfolk naturalist more than three-quarters of a century ago, that the desire to possess "something pretty in a glass case" caused the killing of very many birds, especially of such as were rare and beautiful, which if allowed to exist in our country would maintain the species and be a constant source of pleasure to all who beheld them. For who, walking by a riverside, does not experience a thrill of delight at the sudden appearance in the field of vision of that living jewel, the shining blue kingfisher!

This is one of the favourites of all who desire to have something pretty in a glass case in the cottage parlour in room of the long-vanished pyramid of wax flowers and fruit. It is, however, not only the common people, the cottager and the village publican who desire to possess such ornaments. You see them also in baronial halls. Many a time on visiting a great house the first thing the owner has drawn my attention to has been his stuffed birds in a glass case: but in the great houses the peregrine, and hobby, and goshawk, and buzzard and harrier are more prized than the kingfisher and other pretty little birds.

The Philistine we know is everywhere and is of all classes.

It is to me a cause of astonishment that these mournful mementoes should be regarded as they appear to be, as objects pleasing to the eye, like pictures and statues, tapestries, and other decorative works of art. The sight of a stuffed bird in a house is revolting to me; it outrages our sense of fitness, and is as detestable as stuffed birds and wings, tails and heads, and beaks of murdered and mutilated birds on women's headgear. "Properly speaking,"

said St George Mivart in his greatest work, "there is no such thing as a dead bird." The life is the bird, and when that has gone out what remains is the case.

These dead empty cases are as much to me as to any naturalist, and I can examine the specimens in a museum cabinet with interest. But the mental attitude is changed at the sight of these same dead empty cases set up in imitation of the living creature; and the more cleverly the stuffer has done his work the more detestable is the result.

It may be that some vague notion of a faint remnant of life lingering in the life-like specimen with glass eyes, is the cause of my hatred of the feathered ornament in a glass case. At all events I have had one experience, to be related here, which has almost made me believe that the idea of a sort of post-mortem life in the stuffed bird is not wholly fanciful.

I will call it:

A DIALOGUE OF THE DEAD (AND STUFFED) Ever since I came the wind has been blowing a gale on this furthermost, lonely, melancholy coast, as if I had got not only to the Land's End, but to the end of the world itself, to the confines of Old Chaos his kingdom, a region where the elements are in everlasting conflict. Two or three times during the afternoon I have resolutely put on my cap and water-proof and gone out to face it, only to be quickly driven in again by the bitter furious blast. Yet it was almost as bad indoors to have to sit and listen by the hour to its ravings. From time to time I get up and look through the window-pane at the few cold grey naked cottages and empty bleak fields, divided by naked grey stone fences, and, beyond the fields, the foam-flecked, colder, greyer, more desolate ocean. Would it be better, I wonder, to fight my way down to those wave-loosened masses of granite by the sea, where I would hear the roar and thunder of the surf instead of this perpetual insane howling and screaming of the wind round the house? I turn from the window with a shiver; a splash of rain hurled against it has blotted the landscape out; I go back once more to my comfortable easy-chair by the fire. Patience! Patience! By and by, I say to myself--I say it many times over--daylight will be gone; then the lamp will be brought in, the curtains drawn, and tea will follow, with buttered toast and other good things. Then the solacing pipe, and thoughts and memories and some pleasant waking drawn to while away the time.

What shall this dream be? Ah, what but the best of all possible dreams on such a day as this--a dream of spring! Somewhere in the sweet west country I shall stand in a wood where beeches grow; and it will be April, near the end of the month, before the leaves are large enough to hide the blue sky and the floating white clouds so far above their tops.

Perhaps I shall sit down on one of the huge root-branches, "coiled like a grey old snake," so as to gaze at ease before me at the cloud of purple-red boughs, and interlacing twigs, sprinkled over with golden buds and silky opening leaves of a fresh brilliant green that has no match on the earth or sea, nor under the earth in the emerald mines. I shall watch the love-flight of the cushat above the wood, mounting higher and higher, then gliding down on motionless dove-coloured wings; and I shall listen to the wood

wren, ever wandering and singing in the tree-tops--singing that same insistent, passionate--passionless strain to which one could listen for ever.

I shall ask for no other song, but there will be other creatures there. Down the tall grey trunk of a beech tree before me a squirrel will slip--down, down nearly to the mossy roots, then pause and remain so motionless as to seem like a squirrel-shaped patch of bright chestnut-red moss or lichen or alga on the grey bark. And on the next tree, but a little distance off, I shall presently catch sight of another listener and watcher--a green woodpecker clinging vertically against the trunk, so still as to look like a bird figure carved in wood and painted green and gold and crimson.

Just when I had got so far with the thought of what my dream was to be, I raised my eyes from the fire and allowed them to rest attentively for the first time on a collection of ornaments crowded together in a niche in the wall at the side of the fireplace.

The ornamental objects one sees in a cottage are as a rule offensive to me, and I have acquired the habit of not seeing them; now I was compelled to look at these. There were photographs, little china vases and cups with boys or cupids, and things of that kind; these I did not regard; my whole attention was directed to a pair of glass-fronted cases and the living creatures in them. They were not really alive, but dead and stuffed and set up in life-like attitudes, and one was a squirrel, the other a green woodpecker. The squirrel with his back to his neighbour sat up on his mossy wood, his bushy tail thrown along his back, his two little hands grasping a hazel-nut, which he was in the act of conveying to his mouth. The green woodpecker was placed vertically against his branch, his side towards his neighbour, his head turned partly round so that he looked directly at him with one eye. That wide-open white glass eye and the whole attitude of the bird, with his wings half open and beak raised, gave him a wonderfully alert look, so that after regarding him fixedly for some time I began to imagine that, despite the old dead dusty look of the feathers, there was something of life still remaining in him and that he really was watching his neighbour with the nut very intently.

Why, of course he was alive--alive and speaking to the squirrel! I could hear him distinctly. The wind outside was madly beating against the house and trying to force its way through the window, and was making a hundred strange noises--little sharp shrill broken sounds that mixed with and filled the pauses between the wailing and shrieking gusts, and somehow the woodpecker was catching these small sounds in his beak and turning them into words.

"Hullo!" he said. "Who are you and what
are you doing there?"

"I'm a squirrel," responded the other. "I've said so over and over again, but you will go on worrying me! My only wish is that I could bring my tail just a little more to the right so as to hide my head and paws altogether from you."

"But you can't. Hullo! squirrel, what are you doing there? You forgot to tell me that."

"I'm eating a nut, confound you! You know it; I've told you ten thousand times. I can't ever get it up quite close enough to bite it and I haven't tasted one for seventeen years. One forgets what a thing tastes like."

"I know. I've been fasting just as long myself.
Never an ant's egg! Hullo! Have you got it up?
How does it taste?"

"Taste! You fool! If I could only move I wouldn't mind the nut; I'd go for you like a shot, and if I could get at you I'd tear you to pieces. I hate you!"

"Why do you hate me, squirrel?"

"More questions! Because you're green and yellow like the woods where I lived. There were beeches and oaks. And because your head is crimson red like the agarics I used to find in the woods in autumn. I used to eat them for fun just because they said they were poisonous and it would kill you to eat them."

"And that's what you died of? Hullo! Why don't you answer me? Where did you find red agarics?

"I've told you, I've told you, I've told you, in Treve woods where I lived, very far from here on the other side of Lostwithiel."

"Treve woods, between the hills away beyond Lostwithiel! Why, squirrel, that's where I lived."

"So I've heard; you have said it every day and every night these seventeen years. I hate you."

"Hullo! Why do you hate me?"

"I always disliked woodpeckers. I remember a pair that made a hole in a beech near the tree my drey was in. I played those two yafflers with their laugh laugh laugh some good tricks, and the best of all was when their young began to come out. One morning when the old birds were away I hid myself in the fork above the hole and waited till they crept out and up close to me, when I suddenly burst out upon them, chattering and flourishing my tail, and they were so terrified they actually lost their hold on the bark and tumbled right down to the ground.

How I enjoyed it!"

"You malicious little red beast! You chattering little red devil! They were my young ones, and I remember what a fright we were in when we came back and saw what had happened. It was lucky we didn't lose one! I shall never speak to you again.

There you may sit trying to eat your nut for another seventeen years, and for a hundred years if this horrible life is going to last so long, but you'll never get another word from me."

"I thought that would touch you, woodpecker!

Ha, ha, ha--who's the yaffler now? What a relief; at last I shall be left to eat my nut in peace and quiet, here in this glass case where they put me."

"Why did they put us here?"

"You are speaking to me! Are the hundred years over so soon?"

"There's no one else--what am I to do? Answer me, why did they put us here? Answer me, little red wretch! I don't mind now what you did--they were not hurt after all. You didn't know what you were doing--you had no young ones of your own."

"Hadn't I indeed! My little ones were there close by in the drey."

"And when they were out of the drey did you teach them to run about in the tree, and jump from one branch to another, and pass from tree to tree?"

"I never saw them leave the drey--I was shot."

"Where was that, squirrel?"

"In the Treve Woods where the big beeches are, beyond Lostwithiel."

"Never! Why, that's just where I lived and was shot, too. Did it hurt you, squirrel?"

"I don't know. I saw a flash and remembered no more until I found myself dead in the man's pocket pressed against some wet soft thing. Did it hurt you?"

"Yes, very much. I fell when he fired and tried to get away, but he chased and caught me and the blood ran out on to his hand. He wiped it off on his coat, then squeezed my sides with his finger and thumb until I was dead, then put me in his pocket. There was some dead warm soft thing in it."

Here there was a break in the talk owing to a momentary lull in the wind. I listened intently, but the shrieking and wailing noises without had ceased and with them the sharp little voices had died away. Then suddenly the wind rose and shrieked again and the talk recommenced.

"Hullo!" said the woodpecker. "Do you see a man sitting by the fire looking at us? He has been staring at us that way all the evening."

"What of it! Everyone who comes into this room and sits by the fire does the same. It's nothing new."

"It is--it is! Listen to me, squirrel. He looks as if he could hear and understand us. That's new, isn't it? And he has a strange look in his eyes. Do you know, I think he is going mad."

"I don't mind, woodpecker. I shouldn't care if he were to run out on to the rocks at the Land's End and cast himself into the sea."

"Nor should I. But just think, if before rashing out to put an end to himself he should, in his raving madness, snatch down our cases from the niche and crush them into the grate with his heel!"

"What do you mean, woodpecker? Could such a thing happen?"

"Yes, if he really is insane, and if he is listening to us, and we are making him worse."

"If I could believe such a thing! I should cease to hate you, woodpecker. No, no, I can't believe it!"

"Just think, old neighbour, to have it end at last!

Burnt up to ashes and smoke--feathers and hair, glass eyes, cottonwool stuffing and all!"

"Never again to hear that everlasting Hullo! To hate you and hate you and tell you a thousand thousand times, only to begin it all over again!"

"To fly up away in the smoke, out out out in the wind and rain!"

"The rain! the rain!"

"The rain from the south-west that made me laugh my loudest! Raining all day, wetting my green feathers, wetting every green leaf in the woods beyond Lostwithiel. Raining until all the stony gullies were filled to overflowing, and the water ran and gurgled and roared until the whole wood was filled with the sound."

"No, no, woodpecker, I can't, I can't believe it!"

"It's true! It's true! Don't you see it coming, squirrel? Look at him! Look at him! Now, now!

At last! At last! At last!"

Suddenly their sharp agitated voices fell to a broken whispering and died into silence. For the wind had lulled again. Looking closely at them I thought I could see a new expression in their immovable glass eyes. It frightened me, I began to be frightened at myself; for it now seemed to me that I really was becoming insane, and I was suddenly seized with a fierce desire to snatch the cases down and crush them into the fire with my heel. To save myself from such a mad act I jumped up, and picking up my candle, hurried upstairs to my bedroom. No sooner did I reach it than the wind was up again, wailing and shrieking louder than ever, and between the gusts there were the murmurings and strange small noises of the wind in the roof, and once more I began to catch the sound of their renewed talk.

"Gone! gone!" they said or seemed to say. "Our last hope! What shall we do, what shall we do?

Years! Years! Years!" Then by and by the tone changed, and there were question and answer.

"When was that, squirrel?" I heard; and then a furious quarrel with curses from the squirrel, and "hullos" and renewed questions from the woodpecker, and memories of their life and death in

Treve Wood, beyond Lostwithiel.

What wonder that, when hours later I fell asleep, I had the most distressing and maddest dreams imaginable!

One dream was that when men die and go to hell, they are sent in large baskets-full to the taxidermists of the establishment, who are highly proficient in the art, and set them up in the most perfect life-like attitudes, with wideawake glass eyes, blue or dark, in their sockets, their hair varnished to preserve its natural colour and glossy appearance. They are placed separately in glass cases to keep them from the dust, and the cases are set up in pairs in niches in the walls of the palace of hell.

The lord of the place takes great pride in these objects; one of his favourite amusements is to sit in his easy-chair in front of a niche to listen by the hour to the endless discussions going on between the two specimens, in which each expresses his virulent but impotent hatred of the other, damning his glass eyes; at the same time relating his own happy life and adventures in the upper sunlit world, how important a person he was in his own parish of borough, and what a gorgeous time he was having when he was unfortunately nabbed by one of the collectors or gamekeepers in his lordship's service.

CHAPTER XV

SELBORNE

(1896)

First impressions of faces are very much to us; vivid and persistent, even long after they have been judged false they will from time to time return to console or mock us. It is much the same with places, for these, too, an ineradicable instinct will have it, are persons. Few in number are the towns and villages which are dear to us, whose memory is always sweet, like that of one we love. Those that wake no emotion, that are remembered much as we remember the faces of a crowd of shop assistants in some emporium we are accustomed to

visit, are many. Still more numerous, perhaps, are the places that actually leave a disagreeable impression on the mind. Probably the reason of this is because most places are approached by railroad. The station, which is seen first, and cannot thereafter be dissociated from the town, is invariably the centre of a chaotic collection of ugly objects and discordant noises, all the more hateful because so familiar. For in coming to a new place we look instinctively for that which is new, and the old, and in themselves unpleasant sights and sounds, at such a moment produce a disheartening, deadening effect on the stranger:--the same clanging, puffing, grinding, gravel-crushing, banging, shrieking noises; the same big unlovely brick and metal structure, the long platform, the confusion of objects and people, the waiting vehicles, and the glittering steel rails stretching away into infinitude, like unburied petrified webs of some gigantic spider of a remote past--webs in which mastodons were caught like flies. Approaching a town from some other direction--riding, driving, or walking--we see it with a clearer truer vision, and take away a better and more lasting image.

Selborne is one of the noted places where pilgrims go that is happily without a station. From whichever side you approach it the place itself, features and expression, is clearly discerned: in other words you see Selborne, and not a brick and metal outwork or mask; not an excrescence, a goitre, which can make even a beautiful countenance appear repulsive. There is a station within a few miles of the village. I approached by a different route, and saw it at the end of a fifteen miles' walk. Rain had begun to fall on the previous evening; and when in the morning I looked from my bedroom window in the wayside inn, where I had passed the night, it was raining still, and everywhere, as far as I could see, broad pools of water were gleaming on the level earth. All day the rain fell steadily from a leaden sky, so low that where there were trees it seemed almost to touch their tops, while the hills, away on my left, appeared like vague masses of cloud that rest on the earth. The road stretched across a level moorland country; it was straight and narrow, but I was compelled to keep to it, since to step aside was to put my feet into water. Mile after mile I trudged on without meeting a soul, where not a house was visible--a still, wet, desolate country with trees and bushes standing in water, unstirred by a breath of wind. Only at long intervals a yellow hammer was heard uttering his thin note; for just as this bird sings in the sultriest weather which silences other voices, so he will utter his monotonous chant on the gloomiest day.

It may be because he sung

The yellow hammer in the rain

that I have long placed Faber among my best-loved minor poets of the past century. He alone among our poets has properly appreciated that the singer who never stops, but, "pleased with his own monotony," shakes off the rain and sings on in a mood of cheerfulness dashed with melancholy:

And there he is within the rain,
And beats and beats his tune again,
Quite happy in himself.

Within the heart of this great shower
He sits, as in a secret bower,
With curtains drawn about him:
And, part in duty, part in mirth,
He beats, as if upon the earth
Rain could not fall without him.

I remember that W. E. Henley once took me severely to task on account of some jeering remarks made about our poet's way of treating the birds and their neglect of so many of our charming singers.

In the course of our correspondence he questioned me about the cirl bunting, that lively singer and pretty first cousin of the yellow hammer; and after I had supplied him with full information, he informed me that it was his intention to write a poem on that bird, and that he would be the first English poet to sing the cirl bunting.

He never wrote that lyric, "part in duty, part in mirth"; he was then near his end.

To return to my walk. At last the aspect of the country changed: in place of brown heath, with gloomy fir and furze, there was cheerful verdure of grass and deciduous trees, and the straight road grew deep and winding, running now between hills, now beside woods, and hop-fields, and pasture lands.

And at length, wet and tired, I reached Selborne--the remote Hampshire village that has so great a fame.

To very many readers a description of the place would seem superfluous. They know it so well, even without having seen it; the little, old-world village at the foot of the long, steep, bank-like hill, or Hanger, clothed to its summit with beech-wood as with a green cloud; the straggling street, the Plestor, or village green, an old tree in the centre, with a bench surrounding its trunk for the elders to rest on of a summer evening. And, close by, the grey immemorial church, with its churchyard, its grand old yew-tree, and, overhead, the bunch of swifts, rushing with jubilant screams round the square tower.

I had not got the book in my knapsack, nor did I need it. Seeing the Selborne swifts, I thought how a century and a quarter ago Gilbert White wrote that the number of birds inhabiting and nesting in the village, summer after summer, was nearly always the same, consisting of about eight pairs. The birds now rushing about over the church were twelve, and I saw no others.

If Gilbert White had never lived, or had never corresponded with Pennant and Daines Barrington, Selborne would have impressed me as a very pleasant village set amidst diversified and beautiful scenery, and I should have long remembered it as one of the most charming spots which I had found in my rambles in southern England. But I thought of White continually.

The village itself, every feature in the surrounding landscape, and every object, living or inanimate, and every sound, became associated in my mind with the thought of the obscure country curate, who was without ambition, and was "a still, quiet man, with no harm in him--no, not a bit,"

as was once said by one of his parishioners. There, at Selborne--to give an altered meaning to a verse of quaint old Nicholas Culpepper--

His image stamped is on every grass.

With a new intense interest I watched the swifts careering through the air, and listened to their shrill screams. It was the same with all the birds, even the commonest--the robin, blue tit, martin, and sparrow. In the evening I stood motionless a long time intently watching a small flock of greenfinches settling to roost in a hazel-hedge. From time to time they became disturbed at my presence, and fluttering up to the topmost twigs, where their forms looked almost black against the pale amber sky, they uttered their long-drawn canary-like note of alarm. At all times a delicate, tender note, now it had something more in it--something from the far past--the thought of one whose memory was interwoven with living forms and sounds.

The strength and persistence of this feeling had a curious effect. It began to seem to me that he who had ceased to five over a century ago, whose *Letters* had been the favourite book of several generations of naturalists, was, albeit dead and gone, in some mysterious way still living. I spent a long time groping about in the long rank grass of the churchyard in search of a memorial; and this, when found, turned out to be a modest-sized headstone, and I had to go down on my knees, and put aside the rank grass that half covered it, just as when we look into a child's face we push back the unkempt hair from its forehead; and on the stone were graved the name, and beneath, "1793," the year of his death.

Happy the nature-lover who, in spite of fame, is allowed to rest, as White rests, pressed upon by no ponderous stone; the sweet influences of sun and rain are not kept from him; even the sound of the wild bird's cry may penetrate to his narrow apartment to gladden his dust!

Perhaps there is some truth in the notion that when a man dies he does not wholly die; that is to say, the earthly yet intelligent part of him, which, being of the earth, cannot ascend; that a residuum of life remains, like a perfume left by some long-vanished, fragrant object; or it may be an emanation from the body at death, which exists thereafter diffused and mixed with the elements, perhaps unconscious and yet responsive, or capable of being

vivified into consciousness and emotions of pleasure by a keenly sympathetic presence. At Selborne this did not seem mere fantasy. Strolling about the village, loitering in the park-like garden of the Wakes, or exploring the Hanger; or when I sat on the bench under the churchyard yew, or went softly through the grass to look again at those two letters graved on the headstone, there was a continual sense of an unseen presence near me. It was like the sensation a man sometimes has when lying still with closed eyes of some one moving softly to his side. I began to think that if that feeling and sensation lasted long enough without diminishing its strength, it would in the end produce something like conviction. And the conviction would imply communion. Furthermore, between the thought that we may come to believe in a thing and belief itself there is practically no difference. I began to speculate as to the subjects about to be discussed by us. The chief one would doubtless relate to the bird life of the district. There are fresh things to be related of the cuckoo; how "wonder has been added to wonder" by observers of that bird since the end of the eighteenth century. And here is a delicate subject to follow--to wit, the hibernation of swallows--yet one by no possibility to be avoided.

It would be something of a disappointment to him to hear it stated, as an established fact, that none of our *hirundines* do winter, fast asleep like dormice, in these islands. But there would be comfort in the succeeding declaration that the old controversy is not quite dead yet--that at least two popular writers on British birds have boldly expressed the belief that some of our supposed migrants do actually "lay up" in the dead season. The deep interest manifested in the subject would be a temptation to dwell on it. I should touch on the discovery made recently by a young English naturalist abroad, that a small species of swallow in a temperate country in the Southern Hemisphere shelters itself under the thick matted grass, and remains torpid during spells of cold weather. We have now a magnificent monograph of the swallows, and it is there stated of the purple martin, an American species, that in some years bitter cold weather succeeds its arrival in early spring in Canada; that at such times the birds take refuge in their nesting holes and lie huddled together in a semi-torpid state, sometimes for a week or ten days, until the return of genial weather, when they revive and appear as full of life and vigour as before. It is said that these and other swallows are possessed of habits and powers of which we have as yet but slight knowledge. Candour would compel me to add that the author of the monograph in question, who is one of the first living ornithologists, is inclined to believe that some swallows in some circumstances do hibernate.

At this I should experience a curious and almost startling sensation, as if the airy hands of my invisible companion had been clapped together, and the clap had been followed by an exclamation--a triumphant "Ah!"

Then there would be much to say concerning the changes in the bird population of Selborne parish, and of the southern counties generally. A few small species--hawfinch, pretty chaps, and gold-crest--were much more common now than in his

day; but a very different and sadder story had to be told of most large birds. Not only had the honey buzzard never returned to nest on the beeches of the Hanger since 1780, but it had continued to decrease everywhere in England and was now extinct. The raven, too, was lost to England as an inland breeder. It could not now be said that "there are bustards on the wide downs near Brighthelmstone,"

nor indeed anywhere in the kingdom.

The South Downs were unchanged, and there were still pretty rides and prospects round Lewes; but he might now make his autumn journey to Ringmer without seeing kites and buzzards, since these had both vanished; nor would he find the chough breeding at Beachy Head, and all along the Sussex coast. It would also be necessary to mention the disappearance of the quail, and the growing scarcity of other once abundant species, such as the stone plover and curlew, and even of the white owl, which no longer inhabited its ancient breeding-place beneath the caves of Selborne Church.

Finally, after discussing these and various other matters which once engaged his attention, also the little book he gave to the world so long ago, there would still remain another subject to be mentioned about which I should feel somewhat shy--namely, the marked difference in manner, perhaps in feeling, between the old and new writers on animal life and nature. The subject would be strange to him. On going into particulars, he would be surprised at the disposition, almost amounting to a passion, of the modern mind to view life and nature in their aesthetic aspects. This new spirit would strike him as something odd and exotic, as if the writers had been first artists or landscape-gardeners, who had, as naturalists, retained the habit of looking for the picturesque. He would further note that we moderns are more emotional than the writers of the past, or, at all events, less reticent. There is no doubt, he would say, that our researches into the kingdom of nature produce in us a wonderful pleasure, unlike in character and perhaps superior to most others; but this feeling, which was indefinable and not to be traced to its source, was probably given to us for a secret gratification. If we are curious to know its significance, might we not regard it as something ancillary to our spiritual natures, as a kind of subsidiary conscience, a private assurance that in all our researches into the wonderful works of creation we are acting in obedience to a tacit command, or, at all events in harmony with the Divine Will?

Ingenious! would be my comment, and possibly to the eighteenth century mind it would have proved satisfactory. There was something to be said in defence of what appeared to him as new and strange in our books and methods. Not easily said, unfortunately; since it was not only the expression that was new, but the outlook, and something in the heart.

We are bound as much as ever to facts; we seek for them more and more diligently, knowing that to break from them is to be carried away by vain imaginations. All the same, facts in themselves are nothing to us: they are important only in their relations to other facts and things--to all things, and the essence of things, material and spiritual.

We are not like children gathering painted shells and pebbles on a beach; but, whether we know it or not, are seeking after something beyond and above knowledge. The wilderness in which we are sojourners is not our home; it is enough that its herbs and roots and wild fruits nourish and give us strength to go onward. Intellectual curiosity, with the gratification of the individual for only purpose, has no place in this scheme of things as we conceive it. Heart and soul are with the brain in all investigation--a truth which some know in rare, beautiful intervals, and others never; but we are all meanwhile busy with our work, like myriads of social insects engaged in raising a structure that was never planned. Perhaps we are not so wholly unconscious of our destinies as were the patient gatherers of facts of a hundred years ago. Even in one brief century the dawn has come nearer--perhaps a faint whiteness in the east has exhilarated us like wine. Undoubtedly we are more conscious of many things, both within and without--of the length and breadth and depth of nature; of a unity which was hardly dreamed of by the naturalists of past ages, a commensalism on earth from which the meanest organism is not excluded. For we are no longer isolated, standing like starry visitors on a mountain-top, surveying life from the outside; but are on a level with and part and parcel of it; and if the mystery of life daily deepens, it is because we view it more closely and with clearer vision. A poet of our age has said that in the meanest floweret we may find "thoughts that do often lie too deep for tears." The poet and prophet is not alone in this; he expresses a feeling common to all of those who, with our wider knowledge, have the passion for nature in their hearts, who go to nature, whether for knowledge or inspiration.

That there should appear in recent literature something of a new spirit, a sympathetic feeling which could not possibly have flourished in a former age, is not to be wondered at, considering all that has happened in the present century to change the current of men's thoughts. For not only has the new knowledge wrought in our minds, but has entered, or is at last entering, into our souls.

Having got so far in my apology, a feeling of despair would all at once overcome me at the thought of the vastness of the subject I had entered upon.

Looking back it seems but a little while since the introduction of that new element into thought, that "fiery leaven" which in the end would "leaven all the hearts of men for ever." But the time was not really so short; the gift had been rejected with scorn and bitterness by the mass of mankind at first; it had taken them years--the years of a generation--to overcome repugnance and resentment, and

to accept it. Even so it had wrought a mighty change, only this had been in the mind; the change in the heart would follow, and it was perhaps early to boast of it. How was I to disclose all this to him?

All that I had spoken was but a brief exordium--a prelude and note of preparation for what should follow--a story immeasurably longer and infinitely more wonderful than that which the Ancient Mariner told to the Wedding Guest. It was an impossible task.

At length, after an interval of silence, to me full of trouble, the expected note of dissent would come.

I had told him, he would say, either too much or not enough. No doubt there had been a very considerable increase of knowledge since his day; nevertheless, judging from something I had said on the hibernation, or torpid condition, of swallows, there was still something to learn with regard to the life and conversation of animals. The change in the character of modern books about nature, of which I had told him, quoting passages--a change in the direction of a more poetic and emotional treatment of the subject--he, looking from a distance, was inclined to regard as merely a literary fashion of the time. That anything so unforeseen had come to pass,--so important as to change the current of thought, to give to men new ideas about the unity of nature and the relation in which we stood towards the inferior creatures,--he could not understand. It should be remembered that the human race had existed some fifty or sixty centuries on the earth, and that since the invention of letters men had recorded their observations. The increase in the body of facts had thus been, on the whole, gradual and continuous. Take the case of the cuckoo.

Aristotle, some two thousand years ago, had given a fairly accurate account of its habits; and yet in very recent years, as I had informed him, new facts relating to the procreant instincts of that singular fowl had come to light.

After a short interval of silence I would become conscious of a change in him, as if a cloud had lifted--of a quiet smile on his, to my earthly eyes, invisible countenance, and he would add: "No, no; you have yourself supplied me with a reason for questioning your views; your statement of them--pardon me for saying it--struck me as somewhat rhapsodical.

I refer to your commendations of my humble history of the Parish of Selborne. It is gratifying to me to hear that this poor little book is still in such good repute, and I have been even more pleased at that idea of modern naturalists, so flattering to my memory, of a pilgrimage to Selborne; but, if so great a change has come over men's minds as you appear to believe, and if they have put some new interpretation on nature, it is certainly curious that I should still have readers."

It would be my turn to smile now--a smile for a smile--and silence would follow. And so, with the dispersal of this little cloud, there would be an end of the colloquy, and each would go his way: one to be re-absorbed into the grey stones and long grass, the ancient yew-tree, the wooded Hanger; the other to pursue his walk to the neighbouring parish of Liss, almost ready to believe as he went that the interview had actually taken place.

It only remains to say that the smile (my smile) would have been at the expense of some modern editors of the famous *Letters*, rather than at that of my interlocutor. They are astonished at Gilbert White's vitality, and cannot find a reason for it.

Why does this "little cockle-shell of a book," as one of them has lately called it, come gaily down to us over a sea full of waves, where so many brave barks have foundered? The style is sweet and clear, but a book cannot live merely because it is well written. It is chock-full of facts; but the facts have been tested and sifted, and all that were worth keeping are to be found incorporated in scores of standard works on natural history. I would humbly suggest that there is no mystery at all about it; that the personality of the author is the principal charm of the *Letters*, for in spite of his modesty and extreme reticence his spirit shines in every page; that the world will not willingly let this small book die, not only because it is small, and well written, and full of interesting matter, but chiefly because it is a very delightful human document.

GutenbergTM License when you share it without charge with others.

1.D. The copyright laws of the place where you are located also govern what you can do with this work. Copyright laws in most countries are in a constant state of change. If you are outside the United States, check the laws of your country in addition to the terms of this agreement before downloading, copying, displaying, performing, distributing or creating derivative works based on this work or any other Project GutenbergTM work. The Foundation makes no representations concerning the copyright status of any work in any country other than the United States.

1.E. Unless you have removed all references to Project Gutenberg:

1.E.1. The following sentence, with active links to, or other immediate access to, the full Project GutenbergTM License must appear prominently whenever any copy of a Project GutenbergTM work (any work on which the phrase "Project Gutenberg" appears, or with which the phrase "Project Gutenberg" is associated) is accessed, displayed, performed, viewed, copied or distributed:

This eBook is for the use of anyone anywhere in the United States and most other parts of the world at no cost and with almost no restrictions whatsoever. You may copy it, give it away or re-use it under the terms of the Project Gutenberg License included with this eBook or online at www.gutenberg.org. If you are not located in the United States, you will have to check the laws of the country where you are located before using this eBook.

1.E.2. If an individual Project GutenbergTM electronic work is derived from texts not protected by U.S. copyright law (does not contain a notice indicating that it is posted with permission of the copyright holder), the work can be copied and distributed to anyone in the United States without paying any fees or charges. If you are redistributing or providing access to a work with the phrase "Project Gutenberg" associated with or appearing on the work, you must comply either with the requirements of paragraphs 1.E.1 through 1.E.7 or obtain permission for the use of the work and the Project GutenbergTM trademark as set forth in paragraphs 1.E.8 or 1.E.9.

1.E.3. If an individual Project GutenbergTM electronic work is posted with the permission of the copyright holder, your use and distribution must comply with both paragraphs 1.E.1 through 1.E.7 and any additional terms imposed by the copyright holder. Additional terms will be linked to the Project GutenbergTM License for all works posted with the permission of the copyright holder found at the beginning of this work.

1.E.4. Do not unlink or detach or remove the full Project GutenbergTM License terms from this work, or any files containing a part of this work or any other work associated with Project GutenbergTM.

1.E.5. Do not copy, display, perform, distribute or redistribute this electronic work, or any part of this electronic work, without prominently displaying the sentence set forth in paragraph 1.E.1 with active links or immediate access to the full terms of the Project GutenbergTM License.

1.E.6. You may convert to and distribute this work in any binary, compressed, marked up, nonproprietary or proprietary form, including any word processing or hypertext form. However, if you provide access to or distribute copies of a Project GutenbergTM work in a format other than "Plain Vanilla ASCII" or other format used in the official version posted on the official Project GutenbergTM website (www.gutenberg.org), you must, at no additional cost, fee or expense to the user, provide a copy, a means of exporting a copy, or a means of obtaining a copy upon request, of the work in its original "Plain Vanilla ASCII" or other form. Any alternate format must include the full Project GutenbergTM License as specified in paragraph 1.E.1.

1.E.7. Do not charge a fee for access to, viewing, displaying, performing, copying or distributing any Project GutenbergTM works unless you comply with paragraph 1.E.8 or 1.E.9.

1.E.8. You may charge a reasonable fee for copies of or providing access to or distributing Project GutenbergTM electronic works provided that:

* You pay a royalty fee of 20% of the gross profits you derive from the use of Project GutenbergTM works calculated using the method you already use to calculate your applicable taxes. The fee is owed to the owner of the Project GutenbergTM trademark, but he has agreed to

donate royalties under this paragraph to the Project Gutenberg Literary Archive Foundation. Royalty payments must be paid within 60 days following each date on which you prepare (or are legally required to prepare) your periodic tax returns. Royalty payments should be clearly marked as such and sent to the Project Gutenberg Literary Archive Foundation at the address specified in Section 4, "Information about donations to the Project Gutenberg Literary Archive Foundation."

* You provide a full refund of any money paid by a user who notifies you in writing (or by e-mail) within 30 days of receipt that s/he does not agree to the terms of the full Project GutenbergTM License. You must require such a user to return or destroy all copies of the works possessed in a physical medium and discontinue all use of and all access to other copies of Project GutenbergTM works.

* You provide, in accordance with paragraph 1.F.3, a full refund of any money paid for a work or a replacement copy, if a defect in the electronic work is discovered and reported to you within 90 days of receipt of the work.

* You comply with all other terms of this agreement for free distribution of Project GutenbergTM works.

1.E.9. If you wish to charge a fee or distribute a Project GutenbergTM electronic work or group of works on different terms than are set forth in this agreement, you must obtain permission in writing from the Project Gutenberg Literary Archive Foundation, the manager of the Project GutenbergTM trademark. Contact the Foundation as set forth in Section 3 below.

1.F.

1.F.1. Project Gutenberg volunteers and employees expend considerable effort to identify, do copyright research on, transcribe and proofread works not protected by U.S. copyright law in creating the Project GutenbergTM collection. Despite these efforts, Project GutenbergTM electronic works, and the medium on which they may be stored, may contain "Defects," such as, but not limited to, incomplete, inaccurate or corrupt data, transcription errors, a copyright or other intellectual property infringement, a defective or damaged disk or other medium, a computer virus, or computer codes that damage or cannot be read by your equipment.

1.F.2. LIMITED WARRANTY, DISCLAIMER OF DAMAGES - Except for the "Right of Replacement or Refund" described in paragraph 1.F.3, the Project Gutenberg Literary Archive Foundation, the owner of the Project GutenbergTM trademark, and any other party distributing a Project GutenbergTM electronic work under this agreement, disclaim all liability to you for damages, costs and expenses, including legal fees.

THE END